CONTRIBUTIONS
TO THE
BOTANY OF INDIA

CONTRIBUTIONS

TO THE

BOTANY OF INDIA.

BY

ROBERT WIGHT, M. D., F. L. S., &C.

GYAN

Published by

Gyan Publishing House
5, Ansari Road
Daryaganj, New Delhi-110002
Phone: 011-47034999, 9811692060
E-mail: books@gyanbooks.com

Distribution Network
gyanbooks.com
India, USA, Canada, UK, Australia, France

ISBN: 978-81-212-8322-9 (PB)
First Published, 1834

2nd Impression 2023

Printed at: Gyan Press, Delhi.

CONTRIBUTIONS TO THE BOTANY OF INDIA
Author: ROBERT WIGHT

CONTRIBUTIONS

TO THE

BOTANY OF INDIA.

BY

ROBERT WIGHT, M. D., F. L. S., &c.

MEMBER OF THE IMP. ACAD. NATURÆ CURIOSORUM,

SURGEON ON THE HON. EAST INDIA COMPANY'S MADRAS ESTABLISHMENT.

LONDON:

PARBURY, ALLEN, & CO.

MDCCCXXXIV.

TO

WILLIAM JACKSON HOOKER, LL. D.

PROFESSOR OF BOTANY IN THE UNIVERSITY OF GLASGOW,

NOT ONLY DISTINGUISHED

AS ONE OF THE MOST ACCOMPLISHED AND LIBERAL BOTANISTS

OF THE PRESENT DAY,

BUT AS THE

ZEALOUS PATRON OF NATURAL SCIENCE IN ALL ITS BRANCHES,

THESE CONTRIBUTIONS

ARE RESPECTFULLY DEDICATED

BY HIS MUCH OBLIGED FRIEND

ROB^T. WIGHT.

No. I.

COMPOSITÆ WIGHTIANÆ

JUXTA WIGHTII HERBARIUM AB AUG. PYR. DE CANDOLLE ENUMERATÆ *.

Trib. I.—VERNONIACEÆ. *Cass.; Less. syn.* 145.

I. MONOSIS. *DC. prod. vol. 5. ined.*

Capitulum uniflorum. Invol. oblongi squamæ imbricatæ obtusæ flore multo breviores. Cor. tubulosa 5-fida. Stigmata (Vernoniæ) longe exserta. Achænium glabrum teretiusculum. Pappus 2–3-serialis, setis rigidulis scabris æqualibus.—Suffrutex? Indicus. Folia alterna breviter petiolata obovato-cuneata acutiuscula vix subdentata supra glabra, subtus cum petiolis et ramis tomentosa. Panicula nuda. Rami apice capitula plurima subumbellata conferta subsessilia gerentes.—Affinis *Eremantho* et *Shawiæ*. Nomen ex *μονωσις*, solitudo.

1. M. *Wightiana* (l. c.)——In Indiæ or. montibus "Neelgherry."—*Wight. cat. n.* 1376.—Conyza Wightiana, *Wall.! cat. n.* 3028 (non 3092).

II. VERNONIA. *Schreb. gen.* 2. *p.* 541; *Less. syn.* 146.

§ 1. Strobocalyx (Blume in litt.) *Squamæ involucri floribus multo breviores obtusiusculæ.*

1. V. *elliptica*, caule scandente ramoso tereti subpubescente, foliis breviter petiolatis exacte ellipticis integerrimis supra glabris subtus villis adpressis sericeis argenteis paniculæ ramosissimæ divaricatæ ramis ultimis capitula 3–5 pauciflora gerentibus, invol. ovati squamis ovatis subacutis glabris.——In Neelgherry.—*Wight. cat. n.* 1377.—Affinis ex habitu V. elæagnifoliæ, *Wall.*, sed in specimine imperfecto caract. gener. non rite perspicui.

§ 2. Euvernonia. *Squamæ adpressæ acutæ inappendiculatæ.*

2. V. *Zeylanica* (Less. in Linnæa, 1829, p. 344.)——In Ind. or.—*Wight. cat. n.* 1378.

This is Burm. Zeyl. t. 21.

* The references to Wight's Catalogue, and the observations in English, printed in a smaller type, have been added by *R. Wight* and *G. A. Walker-Arnott*.

3. V. *pectiniformis,* fruticosa?, ramis teretibus glabriusculis, junioribus angulatis pubescentibus, foliis breviter petiolatis lanceolatis acuminatis pectinato acute et elongato—serratis membranaceis supra glabris subtus pilosiusculis, corymbo terminali nudo, capitulis longe pedicellatis multifloris ovatis, invol. squamis siccis glabris ciliatis ovato-oblongis subacutis.——In montibus Dindigul.—*Wight. cat. n.* 1379.—Affinis V. pectinatæ, *DC.* (Eupatorio pectinato, *Wall.*) et forte mera varietas.

Rheed. Mal. 10. t. 62. seems referable to this.

§ 3. TEPHRODES (*Isomeria,* Don in Wight herb.) *Invol. squamæ lanceolatæ lineares acuminatæ subtrinerviæ demum reflexo-patentes. Pappus ext. brevissimus regularis.*

4. V. *physalifolia* (DC. prod. vol. 5.:) caule herbaceo erecto ramoso striato, foliis petiolatis late ovatis repando-dentatis obtusis supra pubescentibus, subtus cauleque cinereo-tomentosis, corymbo laxo, capitulis pedicellatis, invol. squamis lanceolato-linearibus dorso villosis vix acutis.——In Ind. or. Madaraspatana.—*Wight. cat. n.* 1380.—Conyza cinerea, *Wall.! cat. n.* 3000, *c.*—Affinis V. cinereæ.

This appears to be *Serratula cinerea,* Roxb. fl. Ind. 3. p. 406; in E. I. C. mus. tab. 1282.

β *glabrescens;* foliis tenuioribus magis glabratis.—Cum priori.—*Wight. cat. n.* 1383.

5. V. *cinerea* (Less. in Linnæa, 1829, p. 291).—*Wight. cat. n.* 1382.—Frequentissima.

Less. in Linnæa, 1831, p. 673, with synon.—Burm. Zeyl. t. 96. f. 1; Rumph. Amb. 6. t. 14. f. 1; Pluk. t. 177. f. 2.

β *rotundifolia;* foliis etiam superioribus subrotundis obtuse repando-dentatis.—Circa Madras et Negapatam frequens ad littora maris.—*Wight. cat. n.* 1383.—Cacalia rotundifolia, *Willd. phyt.* 2. *p.* 6. *t.* 1; *sp.* 3. *p.* 1732.—Conyza cinerea, *Wall.! cat. n.* 3008, *d.*

This seems to be Rheed. Mal. 10. t. 64.

γ *glabriuscula;* tota fere glabra, caule striato, ramis elongatis erectis, foliis omnibus oblongis, corymbis confertis.——Ad Madras et Negapatam in apricis.—*Wight. cat. n.* 1384.

δ *undulata* (an species propria?) tota cinereo-pubens, caulibus virgatis, foliis lanceolatis dentato-crispis.——In Ind. or.—*Wight. cat. n.* 1385.—Conyza cinerascens, *Wall.! cat. n.* 3062.

Synantherea, Wall.! L. n. 7137, is var. δ.

6. V. *leptophylla* (DC. prod. 5.:) caule herbaceo erecto ramoso tereti tenuiter pubescente, foliis oblongo-linearibus utrinque acuminatis integerrimis tenuissimè pulverulento-scabris, panicula ramosissima laxa divaricata nuda, invol. squamis circ. biseriatis linearibus acuminatissimis vix puberulis.——In Ind. or..—*Wight. cat. n.* 1386.—Conyza linifolia, *Wall.! cat. n.* 3063 (non *Lin.*)—Non male refert. icon. Rumph. 6. t. 14. f. 2. et ideo forte V. fasciculata, *Blum. Bijdr. p.* 893, non *Michx.*

7. V. *conyzoides* (DC. prod. 5.:) caule herbaceo erecto striato brevissimè pubescente, foliis ovato aut oblongo-lanceolatis acuminatis in petiolum brevem attenuatis serratis supra glabris subtus pubescenti-villosis, corymbo composito polycephalo nudo, invol. squamis lanceolato-linearibus pubescenti-villosis acuminatis.——Ad Neelgherry.—*Wight. cat. n.* 1387.—Conyza ovata, *Wall.! cat. n.* 3097 (non *Less.*)——Ad Neelgherry.

8. V. *albicans* (DC. prod. 5.:) tota tomento adpresso sericeo villosa canescens, foliis ovato-rhombeis basi cuneatis integerrimis superne grosse dentatis

acuminatis corymbo subpaniculato nudo, invol. squamis linearibus villosis acuminatis.——In Ind. or.—*Wight. cat. n.* 1388.—Conyza albicans, *Wall.! cat. n.* 3072.—Corollæ lobi hispidi.

§ 4. ASCARICIDA (Cass.). *Squamæ foliaceæ aut in appendicem foliaceum expansæ.*

9. V. *anthelminthica* (Willd. sp. 3. p. 1634).——In ruderatis per omnem Peninsulam frequens.—*Wight. cat. n.* 1389.

This is *V. anthelmintica*, Wall. L. n. 2930.—*Serratula anthelmintica*, Roxb. fl. Ind. 3. p. 405; in E. I. C. mus. tab. 419.—Rheed. Mal. 2. t. 24; Burm. Zeyl. t. 95.

III. CYANOPIS. *Blume in litt.* 1831.—Cyanthyllium, *Blume bijdr.* 889.—Isonema, *Cass. bull. philom.* 1817, non *Brown*, 1810.

Omnia Vernoniæ sed pappus 1-serialis, achænium prismatico-pentagonum.

1. C. *erigeroides* (DC. prod. 5.:) caule herbaceo erecto divaricato ramoso striato glabriusculo, foliis subsessilibus oblongo-linearibus utrinque acuminatis integerrimis glabriusculis, panicula diffusa, capitulis pedicellatis, invol. squamis linearibus acuminatis subpubescentibus disco brevioribus.—Adest culta in h. Wight et spontanea ex ins. Timor. in h. DC.—*Wight. cat. n.* 1390.

IV. DECANEURUM. *DC. prod.* 5.

Omnia Vernoniæ sed pappus 1-serialis et achænium decem-costatum (unde nomen).

§ 1. PHYLLOCEPHALUM (Blume bijdr. 888). *Capitula bracteis foliaceis amplis arcté cincta. Achænia basi non sensim attenuata.*

1. D. *reticulatum*, caule suffruticoso? erecto ramoso setis undique scabro, foliis sessilibus ovatis mucronatis et mucronato-subdentatis supra asperis subtus albo tomentosis nervis venisque scabris reticulatis, capitulis ramos terminantibus, bracteis pluribus foliaceis arcté cinctis, invol. squamis internis chartaceo-scariosis glabris bracteas superantibus.——Ad Neelgherry.—*Wight. cat. n.* 1391.

This is *Amphirephis Indica*, Less. in Linnæa, 6. p. 686; Wall. L. n. 2956.

2. D. *scabridum*, caule herbaceo erecto hinc inde setoso scabro apice tomentoso, foliis in petiolum attenuatis ovato-lanceolatis acuminatis grosse et inæqualiter serratis supra setoso-scabris subtus albo-tomentosis, pedunculis 1-cephalis, bracteis foliaceis capitulo approximatis oblongis subtus tomentosis, invol. squamis internis chartaceo-scariosis glabris.——In India or.—*Wight. cat. n.* 1392.

3. D. *epilejum*, caule herbaceo erecto tereti glabriusculo apice summo tomentoso, foliis in petiolum attenuatis lanceolatis acuminatis grosse serratis supra glabris subtus albo-tomentosis, pedunculis 1-cephalis elongatis, bracteis sub capitulo obtusiusculis patulis valde approximatis subtus tomentosis, invol. squamis internis chartaceo-scariosis glabris subacutis.——In distr. Carnatic ad Courtallum.—*Wight. cat. n.* 1393.—Priori valde affine.

This is *Amphirephis mollis*, Wall. ! L. n. 2957, and perhaps Rheed. Mal. 10. t. 63.

§ 2. GYMNANTHEMUM (Cass. bull. philom. 1817, dict. 20. p. 108). *Capitula aut omnina ebracteata aut rarius bracteolis subulatis inv. squamis inferioribus conformibus instructa. Achænia basi attenuata.*

4. D. *Dendigulense*, collo tuberoso suffruticuloso, caule herbacco erecto parce ramoso tereti basi striato pubescente apice tomentoso, foliis subsessili-

bus late ovatis subrotundisve coriaceis dentatis supra glabris asperulis subtus dense albo-tomentosis summis ovalibus acutis parvis, corymbo fastigiato composito, invol. ovati squamis ovalibus mucronatis imbricatis extus tomentoso-lanatis intus glaberrimis.——In montibus Dindigul.—*Wight. cat. n.* 1394. —Conyza indica, *Wall.! cat. n.* 3042 (non *Blum.*)—Species in genere suo valde distincta. Achænium glabrum inter nervos glandulosum.

5. D. *divergens* (DC. prod. 5.:) caule herbaceo erecto tereti tenuissimè velutino scabro paniculato-ramoso, foliis brevissime petiolatis ellipticis utrinque acuminatis dentatis supra glabris subtus puberulis reticulatis subrugosis, paniculæ ramis elongatis divergentibus foliosis ipsis paniculatis, capitulis ad apices ramulorum confertis, involucr. squamis oblongis acute mucronatis glabriusculis.——In Ind. or.—In Pegu (Belangor.)—*Wight. cat. n.* 1395.—Conyza divergens, *Wall.! cat. n.* 3027 (*excl. litt. e*).—Flores in invol. 7–8. Affine D. aspero (Vernoniæ asperæ, *Hamilt.*)

Wall. L. n. 3027, *e*, is Roxburgh's specimen, cultivated from seeds sent from Mysore: we do not know how it differs; the description (fl. Ind. 3. p. 414), and drawing (in E. I. C. mus. tab. 1637), appear to us referable to a luxuriant cultivated state of our plant: if Roxburgh's be really distinct, the specific name *divergens* ought in justice to be applied to it.

V. ELEPHANTOPUS. *Cass. dict.* 14. *p.* 341.—Elephantopi sp. *Lin.*

1. E. *scaber* (Lin. sp. 1313, excl. syn. Gron. et Browne).——In dumetis ad Courtallum, rarius.—*Wight. cat. n.* 1396.

This is *E. scaber*, Wall. L. n. 2979; Roxb. fl. Ind. 3. p. 445; in E. I. C. mus. tab. 1287.—Rheed. Mal. 10. t. 7.

TRIB. II.—EUPATORIACEÆ. *Less. syn.* 154.

I. AGERATUM. *Lin. gen. n.* 936.

1. A. *conyzoides* (Lin. sp. 1175:) var. *cordifolium*; caule hispido, petiolis hispidissimis foliis late ovatis (infer. forte cordatis?) acuminatis.—*Wight. cat. n.* 1397.—Ageratum cordifolium, *Roxb.! hort. calc. ex Wall.! cat. n.* 3177.—Frequens.

This is *A. cordifolium*, Roxb. fl. Ind. 3. p. 415; *Serratula cordifolia*, Roxb. in E. I. C. mus. tab. 1283.

II. ADENOSTEMMA. *Forst. nov. gen.* (1776), t. 45.—Lavenia, *Swartz, prod.* (1788), *p.* 112.

Ageratum aquaticum, Roxb. fl. Ind. 3. p. 416, or *Ethulia aquatica*, Roxb. in E. I. C. mus. t. 677, belongs to the second section of this genus.

* *Achæniis muricatis.*

1. A. *rugosum* (Wight in Wall.! cat. n. 3221, sub Lavenia:) caule erecto basi tetragono ramoso puberulo seu scabrido, foliis petiolatis late ovatis subcordatis irregulariter (casu?) subtrilobis grosse dentatis nervis subtus elevatis reticulato-rugosis rigidis supra glabriusculis subtus puberulo-scabridis, panicula laxa divaricata polycephala, invol. squamis oblongis dorso scabridis, achæniis muricato-tuberculatis.——In India or.—*Wight. cat. n.* 1398.—Affine Ad. reticulato (in h. Wight observato) sed differt achæniis superficie muricata sed variante et tunc mera varietas esset? (v. s. in coll. Wall.)

** *Achæniis lævibus.*

2. A. *reticulatum*, caule erecto subtetragono glabro secus paniculæ divaricatissimæ ramos parce glanduloso-pubero, foliis (superioribus) ovatis grosse dentatis nervis subtus prominulis reticulatis puberulis rigidis, invol. squamis oblongis obtusis dorso puberulis, achæniis lævibus.——Circa Dindigul.—

Wight. cat. n. 1399.—Folia inferiora non vidi. Valdé accedit ad speciem priorem, an varietas?

3. A. *Madurense* (Wight! sub Lavenia:) caule erecto juniore cum foliis brevissime puberulo adulto glabro, foliis petiolatis late ovato-rhombeis utrinque acuminatis a basi grosse et acute dentatis, corymbi supra ultima folia breviter exserti, ramis puberulis invol. squamis glabris subacutis, achæniis læviusculis.——Prope Madura.—*Wight. cat. n.* 1400.—Hæc species non male refert icon. Burm. Zeyl. t. 42, sed corymbus minus exsertus et folia acute nec obtuse dentata.

4. A. *leiocarpum*, caule erecto tetragono glabro secus paniculæ ramos glanduloso-puberulo, foliis ovato-oblongis subobtusis basi attenuatis subpetiolatis grosse dentatis glabris, panicula laxa polycephala, invol. squamis lineari-oblongis obtusis dorso vix puberulis, achæniis lævibus.——In Ind. or. locis montosis provinciæ Madurensis.—*Wight. cat. n.* 1401.—Lavenia affinis Madurensi, *Wight! herb.*—Herba fere A. elati, fructus A. Mauritiani.

Dr Wight's specimen, examined by De Candolle, lies before us: there are glands, which sometimes become minute tubercles, on the fruit; nor can we distinguish it from *Ad. viscosum*, Forst. (*Lavenia erecta*, Wall. L. n. 3218, *d*), of which Burm. Zeyl. t. 42, appears to us a very tolerable representation.

N.B.—Species hujus generis omnes ulteriori examine egent.

TRIB. III.—ASTEROIDEÆ. *Less. syn.* 161.

I. ASTER. *Linn. gen.*

1. A. *nitidulus*, glaber, caule erecto tereti apice corymboso-ramòso, foliis sessilibus subsemiamplexicaulibus lanceolato-linearibus acuminatis uninerviis margine serrulato-scabris, invol. squamis linearibus acuminatis margine scariosis disco paulo brevioribus.——Neelgherry?—*Wight. cat. n.* 1402.—Ligulæ lineares ex sicco cœruleo-purpureæ styliferæ steriles. Achænia disci compressa glabra tenuia. Pappus rufus 1-serialis sæpe crispatus. Si Galatella rite inter genera admittenda, tunc hæc species ad hoc genus forte amendanda?

II. ERIGERON. *Less. syn.* 190.

§ 3. TRIMORPHÆA (Cass. dict. 37. p. 462.)—*An genus proprium?*

1. E. *sublyratum* (Roxb. ined. ex Don in Wight! herb.:) caule adscendenti-erecto molliter hispido subsimplici, foliis utrinque villosis obovatis grosse inciso-dentatis (dentibus obtusis mucronatis) inferioribus basi attenuatis incisis sublyratis summis cordato-amplexicaulibus, corymbo terminali 7–9-cephalo, capitulis pedicellatis, invol. squamis linearibus acuminatis glabriusculis disco subæqualibus.——In arenosis et apricis provinciæ Tanjore.—*Wight. cat. n.* 1403.—Conyza hirsuta, *Wall.! cat. n.* 3007 (non *Linn.*)—Conyza lacera, *Burm. herb.*—Ligulæ lineares albæ. Achænia glabra elliptica compressa.

Roxburgh's *E. sublyratum* (in E. I. C. mus. tab. 432), an old manuscript name, is the *E. asteroides* of the fl. Ind. 3. p. 435, or Wall. L. n. 2955; consequently, either the present is not Roxburgh's plant (although the drawing referred to be a tolerable representation of our species), or Wall. L. n. 2975, ought also to be referred here.

2. E. *hispidum*, caule basi suffruticuloso erecto ramoso hispido, foliis oblongis hinc inde grosse dentatis subpuberulis infer. obtusis basi attenuatis super. acutis basi dilatatis cordato-amplexicaulibus, capitulis ad apices ramorum 1–2, invol. squamis linearibus acuminatis.——In arenosis agri Maderaspatani.—*Wight. cat. n.* 1404.—Erigeron hirsutum, *Wall.! cat. n.* 2976 (non *Lour.* nec *Pursh.*).—Conyza graveolens, *Wall.! cat. n.* 3106.—Aster Bengalensis, *Heyn.! herb.*—Ligulæ lineares elongatæ 1–2-seriales, fl. fem. multiseriales tubulosi vix subligulati, centr. hermaphr.

3. E. ***multicaule*** (Wall.! cat. n. 3286).

β *Madurense,* caule foliisque pube brevi sparsa scabridis foliis superioribus integerrimis, capitulis omnibus solitariis.——In prov. Madurensi. —*Wight, cat. n.* 1405.—Erigeron multicaule, *Wall.! cat. n.* 3286. *a.*

III. POLOA. *DC. prod.* 5. *ined.*

Capitulum multiflorum heterogamum, fl. radii 1-serialibus fœmineo-sterilibus neutrisve angustissime ligulatis, centralibus tubulosis 5-dentatis hermaphroditis. Rec. planum nudum alveolato-subfimbrilliferum. Invol. pluriseriale, squamis ext. foliaceis oblongis majoribus, inter. subscariosis acuminatis. Anth. et stigm. Asterearum. Achænia teretia glabra erostria. Pappus duplex, exter. brevis coroniformis vix dentatus, in radio magis conspicuus, int. longus setis filiformibus scabris, in radio paucioribus.—Herba indica erecta ramosa glabriuscula aut parce puberula. Folia alterna sessilia oblonga linearia subserrata breviter mucronata. Rami basi foliosi, apice compressi 1-cephali.—Differt ab *Erigeronte* et affinibus pappo duplici, achænio tereti, etc.; a *Pulicaria* antheris ecaudatis, &c.—Genus dicatum, cl. peregrinatori Venetiano Marco Polo qui primus inter hodiernos seculo xiii° Asiam fere totam peragravit et descripsit.

1. P. ***Wightiana.***—Ad Dindigul.—*Wight. cat. n.* 1406.—Erigeron Wightianum, *Wall.! cat. n.* 3284.—Erigeron dubium, *Spreng. in litt. ad Wall.*—Variat glabriusculum aut pubescens.

IV. PSIADIA. *Jacq. hort. Schœnbr.* 3. *p.* 13. *t.* 152.—Psiadia *et* Elphegea, *Cass.* non *Less.*

1. P. ***integerrima*** (DC. prod. 5.)—*Wight. cat. n.* 1407.—Baccharis viscosa, *Lam. dict.* 1. *p.* 345.—Elphegea minor, *Cass. dict.* 14. *p.* 362.——Certe in insulis Mauritianis spontanea, in Peninsula Indica culta.

2. P. ***balsamica*** (DC. prod. 5.)—*Wight. cat. n.* 1408.—Conyza balsamica, *Wall.! cat. n.* 3068.—Baccharis arguta, *Pers. ench.* 2. *p.* 423.——Certe Mauritiana; in India culta.

To this belongs *Eupatorium flexuosum,* Roxb. fl. Ind. 3. p. 415.

V. MYRIACTIS. *Less. in Linnæa,* 1831, *p.* 127; *Syn.* 193.

1. M. ***Wightii,*** erecta, sparse pilosa, foliis inferioribus ovatis basi longe cuneato-attenuatis grosse inciso-dentatis superioribus oblongis integris sessilibus, dentibus apicibusque foliorum calloso-mucronatis.——Ad Neelgherry.—*Wight. cat. n.* 1409.—Valde accedit ad *M. javanicam,* DC. (*Bellidem javanicam,* Blum.! bijdr. 917), et fere utraque media inter *Myriactidem* et *Lagenophoram* ob ligulas bi nec multiseriatas et achænia fere omnino erostria. Involucrum hirsutiusculum post anthesin reflexum. Ligulæ ex sicco albidæ.

VI. VICOA. *Cass. ann. sc. nat.* 1829, tom. 17. *p.* 418; *Less. syn.* 199.

1. V. ***Indica,*** foliis basi auriculatis lanceolatis acuminatis serratis subintegerrimisve utrinque plus minus puberulis, ligulis disco duplo longioribus.——In India or. frequens.—Inula Indica, *Willd. sp.* 3. *p.* 2092; *Wall.! cat. n.* 2964.

This is *Doronicum calcaratum,* Roxb. fl. Ind. 3. p. 434; in E. I. C. mus. tab. 433.—*Aster Indicus,* Willd. sp. 3. p. 2041 (excl. syn. Linn.)—Burm. Zeyl. t. 55. f. 2; Pluk. t. 149. f. 3; t. 340. fol. 29. pl. 13.

α. *serrata;* foliis serratis.—*Wight. cat. n.* 1411.

VII. CÆSULIA. *Roxb. corom.* 1. *p.* 64. *t.* 93.

1. C. *axillaris* (Roxb.! l. c.)——In oryzetis prope Tanjore.—*Wight. cat. n.* 1411; *Wall.! cat. n.* 3183.——Meyera orientalis, *Don, prod.* 180.

This is also *C. axillaris*, Roxb. fl. Ind. 3. p. 447.

VIII. DICHROCEPHALA. *L'her. mss.* 1788.—Centipeda, *Less. syn.* 200, non *Lour.*

Car. gen. Lessingii omnino servandus.

1. D. *latifolia* (DC. prod. 5.)——In Mysore, *Wight. cat. n.* 1412.—Cotula bicolor, *Roth. cat. bot.* 2. *p.* 116.—Cotula latifolia, *Pers. ench.* 2. *p.* 464.—Grangea latifolia, *Lam. ill. t.* 699. *f.* 1.—Cotula sinapifolia, *Wall.! cat. n.* 3237 (*excl. litt. g quæ genus proprium format:* Thespis erecta, *DC.*)—Spilanthus atriplicifolius, *Houtt.! in h. Deless.*

This is *Cotula sinapifolia*, Roxb. fl. Ind. 3. p. 437; in E. I. C. mus. tab. 1643.

2. D. *chrysanthemifolia* (DC. prod. 5.)——In India or.—*Wight. cat. n.* 1413.—Cotula chrysanthemifolia, *Blum.! bijdr.* 918.

IX. SPHÆRANTHUS. *Linn. gen. n.* 998 (non *Scop.*)

§ 1. EUSPHÆRANTHUS. (DC. prod. 5.)

Glomeruli axis plena nec lacunosa. Bracteæ omnes lineares molles capitulis subæquales, exteriores vix ac ne vix involucrum generale constituentes.

1. S. *hirtus* (Willd. sp. 3. p. 2395.)——Frequens in oryzetis peninsulæ Indicæ.—*Blum.! bijdr.* 891; *Wall.! cat. n.* 3178. *a, e, f*; *Wight. cat. n.* 1414.—*Burm. Zeyl. t.* 94. *f.* 3?

To this belongs *S. mollis*, Roxb. fl. Ind. 3. p. 446; in E. I. C. mus. tab. 437; and Rheed. Mal. 10. t. 43. Dr Wight possesses another species, which, although sent to Dr Wallich, is omitted by De Candolle: it is—

S. *Indicus* (Linn.)—S. Indicus, *herb. Wight. olim*; *Wight. cat. n.* 1415.—S. Africanus, *Wall. cat. n.* 3179. *c.*—*Pluk. t.* 108. *f.* 7. (pedunc. passim. delin.); *t.* 312. *f.* 6.

§ 2. POLYCEPHALOS. (Forsk. descr. Æg. p. 154.)—*Oligolepis*, Cass. dict. 50. p. 202.

Glomeruli axis cavo-lacunosa. Bracteæ omnes ovato-lanceolatæ cuspidatæ rigidæ capitala superantes, exteriores vix ac ne vix invol. gener. constituentes.

2. S. *amaranthoides* (Burm. fl. Ind. 186.)——In humidis provinciæ Tanjore ad oppidum Negapatam.—*Wight. cat. n.* 1416.—S. Zeylanicus, *Heyn. ex Wall. cat. n.* 3180.—*Pluk. amalth. t.* 357. *f.* 6.—Affinis *S. suaveolenti*, sed differt foliis angustioribus interdum integerrimis subtus punctatis, pedunculis brevissimis apteris, bracteis acuminatissimis.

X. BLEPHARISPERMUM. *Wight.*

Capitula innumera in glomerulum globosum coacervata, biflora: fl. tubulosis; altero fœmineo breviter 3–4-dentato gracili; altero masculo 5-dentato. Recept. capituli angustissimum, palea lineari inter flores interjecta. Invol. squamæ exter. paucæ (2?), breves, ovatæ, inter. 2 flosculis æquales, altera sub flore fœmineo plana, alt. sub flore masculo conduplicata. Stylus brevis inclusus, apice vix bifidus, infra emarginaturam tumidulus subpuberulus. Antheræ apice appendiculatæ basi subcaudatæ. Achænium fl. fem. obcompressissimum obovali-oblongum subtetragonum, angulis duobus nerviformibus glabris, 2 marginalibus longe et regulariter ciliatis!: fl. masc. ovarium gracile elongatum glabrum. Pappus ach. fœm. paleis scariosis 2–5 acuminatissimis, ach. masc. nullis.—Suffrutices Ceylonenses glaberrimi. Rami teretes graciles. Folia alterna ovata aut ovalia integerrima penninervia. Pedunculi 1–3 ad apicem rami 1-cephali.—An potius capitula 4-flora dicenda? nempe colligens dua contigua in unicum? Car. præcipue ex *Cl. petiolari*.—Genus singularissimum ad Cynareos primo ex stylo relatum, sed potius accedere videtur ad *Sphæranthum* mediante *Athroismate*, DC. (*Sphærantho laciniato*, Wall. cat. n. 3184.)

*1. B. *petiolare*, foliis petiolatis ovato-lanceolatis acuminatis, glomerulis ternis longe pedunculatis, bracteis invol. universalis parvis membranaceis capitulis brevioribus.——In Ceylona prope Ugandamalej.—*Wight. cat. n.* 1417.

To this, perhaps, ought to be referred Pluk. t. 391. f. 1. The following is the description made by Klein when he collected the plant, Feb. 29. 1796.—"*Syngenesia Polyg. Segregata.*—*Calyx communis* polyphyllus; foliolis ovato-lanceolatis. *Perianthium partiale* 4-florum pentaphyllum; foliolis ovatis carinatis, quinto minimo, duobus hermaphroditos, duobus fœmineos includentibus.—*Corolla composita* ex hermaphrodis duobus et fœmineis duobus. *Corolla propria* hermaphroditi monopetala, infundibuliformis 5-fida. *Corolla* fœminea tubulosa, tridendata.—*Stamina.* Filamenta 5, brevissima. Anthera cylindracea, tubulosa.—*Pistillum* hermaphroditis: germen obsoletum: stylus filiformis: stigma obtusum, prominulum. Fœmineis: germen ovatum, compressum: stylus filiformis: stigma bifidum.—Pericarpium nullum.—*Semen hermaphroditis* sterile: *fœmineis* ovatum, compressum, e lateribus ciliatum. Pappus coronatus e setis duobus emergentibus inæqualibus.—*Receptaculum commune* squamosum; *partiale* paleaceum.—*Caulis* suffruticosus, teres.—*Folia* alterna, petiolata, ovato-lanceolata, integerrima, glabra.—*Flores* pedunculati, capitati, terminales axillaresque."

2. B. *subsessile*, foliis ellipticis utrinque attenuatis subsessilibus, gomerulo terminali solitario subsessili, bracteis invol. univ. foliaceis capitula superantibus.——In Ceylona.—*Wight. cat. n.* 1418.—*Wall. cat. n.* 7524 (sine nomine).

XI. GRANGEA. *Adans. fam. p.* 121.

1. G. *Maderaspatana* (Poir, suppl. 3. p. 825. *non* Desf. Delil.)——In pascuis humidis prope Madras frequens.—*Wight. cat. n.* 1419.—Cotula Maderaspatana, *Willd. sp.* 3. *p.* 2170; *Wall. cat. n.* 3235.—Grangea Adansonii, *Cass. dict.* 19. *p.* 304.—Species Ægyptiaca (*G. ægyptiaca*, DC. prodr. 5) cum hac a pluribus confusa est diversa præsertim ex fructus structura.

This is *Artemisia Madraspatana*, Roxb. fl. Ind. 3. p. 422; in E. I. C. mus. tab. 424.—Rheed. Mal. 10. t. 49.

XII. CYATHOCLINE. *Cass. an. sc. nat.* 1829. *tom.* 17. *p.* 419.

1. C. *lyrata* (Cass. l. c.)——In Ind. or. variis locis; in Pegu (Belanger!), Nepaulia (Wall.!), etc.—*Wight. cat. n.* 1420.—Tanacetum viscosum, *Wall.! cat. n.* 3232, *a, b, c, d,* (excl. jam monente cl. Lessing, litt. *e, f,* ad *Cyathoclinem strictam*, DC. referendis).

Part of Dr Wight's specimens are from Rottler, given him by Heyne, and collected in the Circars: they bear the name of *Artemisia paniculata;* these we presume to be Wall. L. n. 3231. *f,* but they are certainly not distinct from the specimens examined by De Candolle, and here called *C. lyrata.* Lessing's characters appear to us only to apply to extreme states, between which our specimens seem to be intermediate.

XIII. BLUMEA. *DC.* (non *Spreng.* nec *Nees.*)—Erigeron, *Don. prod. fl. Nep.* 171. (non *Linn.*)—Conyzæ spec. *peræque auct. Ind.*

Capitulum multiflorum heterogamum, fl. radii multiserialibus fœmineis tenuissime tubulosis truncatis aut 2–3-dentatis, disci paucis (5–25) masculis tubulosis 5-dentatis. Rec. nudum planum. Inv. pluriseriale imbricatum, squamis linearibus acuminatis. Antheræ basi ecaudatæ. Achænium teretiusculum (nec ut in Conyza et Eregeronte compressum). Pappus 1-serialis, setis capillaribus vix scabris.—Herbæ Indicæ (paucissimæ Africanæ) interdum suffrutescentes. Folia alterna. Capitula sæpius corymbosa.—Genus Indicum numero specierum (fere 80) insigne dicatum cl. Blume de Botanica Indica optime merito et inter Botanicos præclaros rite numerando. *Blumea* Spreng. est *Reinwardtia* Blume, et *Blumea* Nees est *Talauma;* unde nomen vacat!

§ 1. APTERÆ, *nempe foliis non decurrentibus.*

1. B. *arenaria* (DC. prod.:) basi glabriuscula apice undique sericeo-villosissima, caule a basi ramosa diffuso tereti, foliis oblongis sessilibus mucrone calloso acutis integerrimis aut acute denticulatis, pedunculis solitariis 1-cephalis hirsuto-lanatis axillaribus terminalibusque, invol. squamis linearibus acuminatis dorso hirsutis apice ciliatis, corollis masculis apice barbatis.——In arenosis prope Negapatam.—*Wight. cat. n.* 1422.—Conyza villosa, *Wall.! cat. n.* 3105 (non 3083, nec *Willd.*).—Radix longa gracilis. Fl. fœm. stylo longe exserto, masc. pauci.

2. B. *tenella* (DC. prod. 5.:) caule herbaceo multicipiti puberulo tereti parce ramoso, foliis oblongis basi obtuse cordato semiamplexicaulibus acutis distanter acute et calloso-dentatis pube parva scabridis summis acuminatis subintegerrimis, pedunculis terminalibus et oppositifoliis 1-cephalis puberis folio triplo longioribus, invol. squamis linearibus acuminatis disco fere sublongioribus.—In arenosis versus Negapatam, et etiam ex herb. DC. in insula Timor.—*Wight cat. n.* 1422.—Fl. flavi, fœm. innumeri, masc. 8–9, antheris inclusis. Affinis B. amplectenti, *DC.* (*Wall. n.* 3096), et B. oligocephalæ (*Wall.! cat. n.* 3104.)

3. B. *amplectens* (DC. prod. 5.:) caule herbaceo piloso tereti divaricato-ramoso, foliis junioribus villosis, adultis subglabratis, caulinis ovatis semiamplexicaulibus acute dentatis, infinis hinc inde basi incisis, pedunculis paucis subterminalibus folio paulo longioribus.—In ruderatis ut arenosis frequens circa Negapatam.—*Wight. cat. n.* 1423.—Erigeron obliquum, *Linn. mant.* 572? (ex *Russ. herb.*, et ideo Conyza obliqua, *Willd. sp.* 3. *p.* 1930?)—Conyza amplexicaulis, *Lam. dict.* 2. *p.* 85?—Conyza amplectens, *Wall! cat. n.* 3096.—Fl. fœm. pluriseriales stigm. exsertis rubris, masc. 15–20 antheris exsertis.

4. B. *anagallidifolia* (DC. prod. 5.:) caulibus a collo multiplicibus diffusis ramosis teretibus apice puberulis, foliis caulinis ovali-oblongis sessilibus hinc inde acute dentatis, rameis lanceolatis integerrimis puberulis, capitulis longe pedicellatis solitariis axillaribus terminalibusque in paniculam foliosam laxam subcorymbosam dispositis, invol. squamis linearibus acutis glabriusculis disco paulo longioribus.——In humidis ad Negapatam frequens, etiam in Pegu ex cl. Belanger.—*Wight. cat. n.* 1424.—Fl. fœm. innumeri, masculi 10–12 cor. apice subbarbatis.—Conyza anagallidifolia, *Wall.! cat. n.* 3098.

Rheed. Mal. 10. t. 18, and Pluk. t. 19. f. 4, probably belong to this species.

5. B. *oligocephala* (DC. prod. 5.:) parce puberula herbacea multicaulis subdiffusa, foliis inferioribus obovato-oblongis basi cuneatis super. oblongo-ova-

libus sessilibus subsemiamplexicaulibus omnibus apice remote et acute dentatis, pedunculis 1–3 ex apice caulium villosis 1-cephalis folio longioribus, invol. squamis linearibus acutis discum æquantibus.——Circa Madras.—Conyza amplexicaulis, *Wight. in Wall. ! cat. n.* 3104 (non *Lam.* nec *Mich.*)—Fl. masc. 15–20 antheris exsertis (descr. ex coll. *Wall.*)

6. B. *bifoliata* (DC. prod. ;) caule herbaceo adscendente tereti ramoso apice præsertim puberulo, foliis acute dentatis infer. ovatis basi longe attenuatis glabriusculis superioribus sessilibus oblongis cuneatis puberis, pedunculis axillaribus supra basin sæpius folia duo subopposita gerentibus, pedicellis elongatis 1-cephalis, invol. squamis linearibus acutis disco subæqualibus.——In nemorosis humidis prov. Tangore, et Travancore, ac in multis aliis locis frequens.—*Wight. cat. n.* 1425.—Conyza foliolosa, *Wight! in Wall.! cat. n.* 3092 (non 3030, nec *Willd.*)—Conyza bracteolata, *Rottl.! in herb. Puer.*—*Pluk. alm. t.* 177. *f.* 1.—Cor. masc. 8–16 flavidæ apice vix barbellatæ, antheris inclusis.

This is *Conyza bifoliata*, Linn. sp. p. 1207 ; Roxb. fl. Ind. 3. p. 430 ; in E. I. C. mus. tab. 431 ; and *C. axillaris*, Lam.

7. B. *barbata* (DC. prodr. :) caule herbaceo erecto simplici tereti pilosiusculo, foliis lanceolatis basi attenuatis acute et distanter serratis acutis supra parce puberis subtus dense et adpresse sericeo-villosis, ramis axillaribus folio brevioribus foliolis apice racemosis, summis corymbosis, invol. squamis linearibus acutis glabriusculis disci longitudine.——In Ind. or.—*Wight cat. n.* 1426.—Conyza barbata, *Wall. ! cat. n.* 3099.—Fl. flavidi, fœm. innumeri stylo exserto bifido, masc. 5 cor. apice glabriusculis.

8. B. *Wightiana* (DC. prod. :) caule herbaceo erecto tereti striato puberulo, foliis inferioribus petiolatis ovalibus basi cuneatis obtuse et inequaliter dentatis subobtusis super. cuneatis ovali-oblongis, capitulis in paniculam racemosam interruptam terminalem digestis, invol. squamis linearibus acutis intimis glabriusculis disco paulo longioribus, corollis masculis extus tuberculoso-glandulosis.——In India prope Madras et ex Belanger prope Pondichery.—*Wight. cat. n.* 1427.—Conyza Wightiana, *Wall. ! cat. n.* 3093.—Fl. fœm. innumeri tenues, masc. 8–12 ex sicco rubentes.

In Dr Wight's specimens, named by De Candolle, the leaves are very acutely toothed. *Conyza bifoliata*, herb. Madr. in Wall. ! L. n. 3091, seems to be quite the same.

9. B. *lacera* (DC. prod.)—*Wight. cat. n.* 1428.—Conyza lacera, *auctor.*

α, *Burmanni.*—*Burm. fl. Ind.* 180. *t.* 59. *f.* 1.

β, *Commersonii.*—C. lacera, *Lam.* ! *dict.* 2. *p.* 83.

γ, *Wallichii ;* foliis obovatis profunde et acute dentatis, panicula laxa.——In Silhet, China, etc.—C. lacera, *Wall. ! cat. n.* 3045 et 3082, forte etiam C. villosa, *Wall. cat. n.* 3083.

δ. *Blumei.*—C. Javanica, *Blume ! Bijdr.* 897.

To this species belongs *C. lacera*, Roxb. fl. Ind. 3. p. 428 ; in E. I. C. mus. tab. 429.

10. B. *glandulosa* (DC. prod. :) tota brevissime piloso-glandulosa, caule herbaceo erecto subtereti striato-nervoso, foliis caulinis breviter petiolatis ovatis subacuminatis irregulariter dentatis basi lyrato-incisis rameis basi cuneatis obovato-oblongis subdentatis, capitulis ad apices ramorum paniculam foliosam constituentibus, pediculis ramosis glandulosis folio brevioribus, involucr. squamis linearibus acutis disco æqualibus.——Prope Negapatam.—*Wight. cat. n.* 1429.—Fl. flavidi glabri, masc. circiter 10. Recep. convexum.

C. bifoliata, herb. Madr. (not Wall. L. n. 3091), forms part of Wight's No. 1429.

11. B. *virens* (DC. prod. :) caule herbaceo erecto tereti glabro ramoso, foliis rigidulis sessilibus utrinque sparse puberulis caulinis elongatis basi cu-

neatis sessilibus trifidis (lobis lanceolatis mucronatis regulariter serratis terminali longiore) rameis obovatis basi cuneatis sparse serratis, panicula longa, pedicellis axillaribus folio longioribus 1–5-cephalis, capitulis pedicellatis, invol. squamis inter. linearibus disco longioribus mucronatis.——In arenosis maritimis ad Negapatam. Circars; *Heyne; Russell.*—*Wight. cat. n.* 1430.—Conyza virens, *Wall.! cat. n.* 3037. *c.*—Erigeron virens, *Russ.! herb.*—Fl. purpurascentes, masc. 5.

12. B. *lapsanoides* (DC. prodr.:) glabra, caule herbaceo erecto tereti ramoso, foliis petiolatis membranaceis caulinis sinuato-repandis hinc inde mucronato-serrulatis (lobis mucronatis lateralibus subtriangularibus terminali ovato) rameis obovatis vix subserratis, panicula laxa oligocephala, ramis lateralibus foliosis, capitulis pedicellatis, invol. squamis linearibus mucronatis disco paulo longioribus.——Ad Courtallum.—*Wight. cat. n.* 1431.—Conyza virens, *Wall.! cat. n.* 3037. *a, b, d.*—Fl. flavidi, masc. 10–12. Achænia glabra. Affinis priori et B. hymenophyllæ (*Wall.! cat. n.* 3038.)

13. B. *Heyneana* (DC. prod.:) caule herbaceo erecto ramosissimo subtilissime striatulo et puberulo, foliis oblongis basi attenuatis apice obtusis hinc inde dentes exsertos acutos gerentibus utrinque glabriusculis infer. basi lyrato-auritis, panicula foliosa laxa ramis folii longitudine apice subcorymbosis, invol. squamis linearibus acuminatis dorse subpuberulis disco paulo longioribus.——In Ind. or.—*Wight. cat n.* 1432.—Conyza Heyneana, *Wall.! cat. n.* 3089.—Conyza aurita, *Heyn.! herb.* (non *Linn. f.*)—Fl. flavidi, fœm. innumeri stylo exserto, masc. circit. 12.

14. B. *hieracifolia* (DC. prod:) tota hirsuta, caule herbaceo erecto tereti simplici, foliis calloso-dentatis infer. obovatis obtusis in petiolum attenuatis super. ovalibus oblongisve acutis sessilibus aut semiamplexicaulibus, capitulis sessilibus confertis in thyrsum ovato-oblongum digestis, inv. squamis linearibus acuminatis glabriusculis disco longioribus.——In Ind. or.—*Wight. cat. n.* 1433.—Erigeron hieracifolium, *Don. prod.* 272.—Conyza communis, *Wall.! cat.* 3018.—Conyza hieracifolia. *Spreng. syst.* 3. *p.* 514.—Fl. fœm. innumeri stylo exserto subindiviso, masc. 5.

15. B. *glomerata* (DC. prod.:) tota villoso-subhirsuta cinerea, caule herbaceo erecto tereti ramoso, ramis in paniculam dispositis foliosis apice capitula sessilia glomerata interrupte spicata gerentibus, foliis lanceolatis serratis utrinque acutis, invol. squamis multiserialibus, ext. minimis, intimis linearibus flores paulo superantibus.——In arenosis maritimis circa Madras et Negapatam. Ad Montes Circars; *Heyne; Roxburgh.*—*Wight. cat. n.* 1434.—Conyza glomerata, *Wall.! cat. n.* 3055.—Fl. fœm. vix masculis numerosiores truncati, stigmate bifido purpureo; masc. circiter 30!, antheris luteis exsertis.

Con. fistulosa, Roxb. Ind. 3. p. 420; *C. spicata*, Roxb. in E. I. C. mus. tab. 428; certainly belongs to this species.

16. B. *eriantha* (DC. prod.:) caule herbaceo erecto tereti basi glabro apice ramoso et mollissimo hispido, foliis membranaceis caulinis sessilibus ovalibus acute et exserte apiculato-dentatis molliter pubescentibus summis cordato-amplexicaulibus, pedunculis axillaribus et terminalibus 1-cephalis solitariis aut paucis aggregatis demum valde elongatis, invol. squamis linearibus ciliatis acuminatis, corollis fœm. et masc. apice barbatis.——In Ind. or.—*Wight. cat. n.* 1435.—Fl. masc. 25–30.

17. B. *oxyodonta* (DC. prod.:) herbacea villis longis sericeis laxe villosa, caulibus pluribus tenuibus simplicibus diffusis, foliis argute et grosse serratis subtus magis sericeo-villosis infimis obovatis petiolatis super ellipticis subsessilibus, capitulis paucis ad apices ramorum racemoso-corymbosis subsessilibus, involucri squamis linearibus acuminatis disco paulo longioribus.——Ad littora maris versus Negapatam.—*Wight. cat. n.* 1436.—Conyza oxyodonta, *Wall.! cat. n.* 3015.—Fl. fœm. pluriseriales, masc. 15–20. Pappus parcus, tenuis, deciduus.

§ 2. CAULOPTERÆ, *nempe foliis secus caulem in alam foliaceam decurrentibus.*

18. B. *pterodonta* (DC. prod.:) caule herbaceo tereti ramoso vix subpuberulo, foliis elliptico-oblongis glabris subserratis in alam profunde et acute dentatam fere incisam longe productis, ramis foliosis subpaniculatis apice mono aut oligo-cephalis, pedicellis nudis, invol. squamis ext. oblongis foliaceis brevibus intimis scariosis linearibus acutis flores paulo superantibus.—— Ad littora maris, locis arenosis magis in Coromandelia frequens.—*Wight. cat. n.* 1437.—Conyza alata, *Wall.! cat. n.* 3039, *b.*—Facile differt a B. alata cujus alæ decurrentes integræ.

Part of Dr Wight's specimens are from Rottler's herbarium, under the name of *C. alata*, obtained from Heyne, and gathered in Mysore: consequently 3039. *a*, of Dr Wallich's List ought surely not to be separated.

19. B. *aurita* (DC. prod.:) hirsuto-villosa, caule erecto tereti ramoso, foliis lyrato-pinnatipartitis (lobis oblongis acutis dentatis terminali majore) infer. basi decurrentibus secus caulem 1–2 auritis, panicula terminali laxa, capitulis pedicellatis erectis invol. squamis lanceolatis acuminatis ext. dorso hirsutis inter. subscariosis disco paulo longioribus.——In ruderatis et apricis circa Madras frequens et per totum littus Coromandelianum.—*Wight. cat. n.* 1438.—Conyza aurita, *Lin. f. suppl.* 367; *Wall.! cat. n.* 3069 et 3086.

Con. aurita, Roxb. fl. Ind. 3. p. 428; *C. terebinthacea*, Roxb. in E. I. C. mus. tab. 430; Pluk. t. 422. f. 1; and perhaps 60. f. 3. (from a bad specimen) are referable here.

XIV. CONYZA. *Less. syn.* 203.—Eschenbachia, *Mœnch. meth.* 573.—Dimorphantes *et* Fimbrillaria, *Cass.*—Conyzæ sp., *auct.*

Caract. Lessingiano adde: achænia compressa.

1. C. *absinthifolia* (Wall.! cat. n. 2994:) pilosiusculo-pubescens erecta ramosissima, foliis obovato-oblongis mucronatis basi cuneato-attenuatis infer. grosse paucidentatis super. integerrimis, panicula corymbosa ramosissima polycephala, invol. squamis linearibus acuminatis disco brevioribus.——In montanis Dindigul et Neelgherry; Vellore, *Rottler.*—*Wight. cat. n.* 1439.—Conyza pinnatifida var. *Roxb. cat. calc.* 61.—Erigeron pinnatifidum var. *Don. prod.* 172.—Ambigit inter Erigerontem et Conyzam ob flores fœmineos extus subproductos et ligulæ originem quasi indicantes. Achænium parvum glabrum lenticulare compressum. Pappus subrufescens.

This is *C. pinnatifida*, Roxb. fl. Ind. 3. p. 430, and *C. stricta*, Willd. sp. 3. p. 1922, according to Willdenow in herb. Madr., Rottl., and Heyne. Perhaps it might be only justice to restore the name *stricta*.

XV. EPALTES. (*Cass. bull. philom.* 1818, *p.* 139.)—Ethulia, *Gærtn.* non *Cass.*

1. E. *divaricata* (Cass. l. c. et dict. 15. p. 6.)——In oryzetis Ind. or. frequens.—*Wight. cat. n.* 1440.—Ethulia divaricata, *Lin. mant.* 110; *Burm. Ind. t.* 58. *f.* 1; *Wall.! cat. n.* 2988. *a, b.*—(*Pluk. t.* 160. *f.* 5; *t.* 378. *f.* 2.)

The figure in Pluk. t. 21. f. 4, usually quoted here, appears to belong to Wall. L. n. 2988, *c*.

XVI. PLUCHEA. *Less. syn.* 207.—Pluchea *et* Chænobolus, *Cass. dict.* 42. *p.* 1. *et* 49. *p.* 337.

NB. Genus polytypum in posterum verisimiliter dividendum!

Conyza corymbosa, Roxb. fl. Ind. 3. p. 426; in E. I. C. mus. tab. 1172; appears to be *P. Indica*, Less.

1. P. *tomentosa:* fruticosa, ramis teretibus, ramulis et foliis utrinque pube brevi vix tomentosa griseo-velutinis, foliis obovatis ovalibusve cordato-semiamplexicaulibus dentatis subdecurrentibus, pedunculis ex axillis superioribus

ortis folio longioribus capitula pedicellata conferta gerentibus in paniculam confertam brevem dispositis, invol. squamis oblongo-linearibus acutis disco paulo brevioribus.——In agro Mysore.—*Wight. cat. n.* 1441.—Conyza tomentosa, *Wall.! cat. n.* 3031 (non *Mill.* nec *Forsk.*)—Eupatorium Mysorense, *Heyn.! herb.*—Affinis Pl. Indicæ, *Less.*, et Pl. Dioscoridis, *DC.*

XVII. SIEGESBECKIA. *Linn. hort. diff.* 412.

1. S. *orientalis* (Linn. sp. 1269.)——In montibus Dindigul altitud. 2000 ad 2500 ped.—*Wight. cat. n.* 1442.

S. orientalis, Roxb. fl. Ind. 3. p. 439; Wall. L. n. 3182.—*S. brachiata*, Roxb. l. c., and in E. I. C. mus. tab. 434, appears to be the same.

XVIII. ECLIPTA. *Linn. mant.* 157.—Micrelium, *Forsk.*

1. E. *erecta* (Linn. mant. 286 ? ; Willd. sp. 3. p. 2217.)——In graminosis humidis et argillosis frequens.—*Wight. cat. n.* 1443.—E. erecta, *Linn.*, est forte E. longifolia, *Schrad.*, sed cum omnes auctores hanc pro specie Linnæana receperint etiam nomen servandum videtur.

"Annual, scabrous, from erect to prostrate. *Verbesina prostrata*, and *alba*, sp. pl. p. 1272, and I suspect *E. erecta*, *punctata*, and *prostrata*, are only one and the same species, or at most but accidental varieties, from age, soil, and situation ; at least I have never found any other in India to compare with them."—ROXB.

2. E. *parviflora* (Wall.! cat. n. 3213 :) caule prostrato repente adpresse strigoso, foliis oblongo-lanceolatis utrinque acuminatis subintegerrimis subscabridis pedunculis 1–2 capitulo vix duplo longioribus.——In humidis frequentissima.—*Wight. cat. n.* 1444.—Habitus E. prostratæ et folia E. erectæ, sed caulibus radicantibus ab utraque diversa.

Almost all the species of the genus throw out roots, in a wet soil, from near the bottom of the stems : the shape of the leaves is very inconstant.—Perhaps Pluk. t. 118. f. 5, belongs to this species.

XIX. BLAINVILLEA. *Cass. journ. phys.* 1823, *p.* 216 ; *dict.* 29. *p.* 493.

1. B. *latifolia* (DC. prod. :) foliis rhomboideis acuminatis grosse dentatis utrinque sparsè pubescenti-hirtellis, ramis (petiolisque) hirtellis oppositis summis dichotomis, pedunculis petiolo plerumque brevioribus.——Apud Rajahmundry, Pathoocottah, ad montes Dindygul et in humidis prov. Madurensis.—*Wight. cat. n.* 1445.—Eclipta latifolia, *Linn. f. suppl.* 378. et *Linn. herb.* (ex *Wight !*).—Verbesina dichotoma, *Wall.! cat. n.* 3204. *c, e,* (non *Murr.* quæ ad *B. dichotomam*, DC. seu *B. rhomboideam*, Cass. referenda).

This is *Verbesina Lavenia*, Roxb. fl. Ind. 3. p. 442. (not *Linn.*) or *V. calendulacea*, Roxb. in E. I. C. mus. tab. 435 (not of authors), and Pluk. t. 382. f. 6.

TRIB. IV.—SENECIONIDEÆ. *Less. syn.* 218.

I. XANTHIUM. *Linn. gen.* 1056.

1. X. *indicum* (Roxb. cat. calc. 67. :) involucro fructigero ovali inter aculeos et ad basin rostrorum pubescente, rostris apice uncinatis.——In ruderatis frequens.—*Wight. cat. n.* 1446.—X. orientale, *Linn. sp.* 1400 (excl. syn.).—X. chinense, *Mill. dict.* ?—X. Indicum, *Wall.! cat. n.* 3181.—Differt a *X. strumario* rostris apice uncinatis nec rectis.

II. WEDELIA. *Jacq.* ; *Less. syn.* 222. (non *Adans.*)

§ 1. *Involucro uniseriali* 4–5 *phyllo.*

1. W. *calendulacea* (Less. syn. 222., non Rich.).——In Ind. or.—*Wight.*

cat. n. 1447.—Verbesina calendulacea, *Linn. sp.* 1272; *Linn. herb.* (ex *Wight!*); *Wall.! cat. n.* 3205.—*Burm. Zeyl. t.* 22. *f.* 1.

This is *Verb. calendulacea*, Roxb. fl. Ind. 3. p. 440; in E. I. C. mus. tab. 978; and Rheed. Mal. 10. t. 42.

§ 2. *Involucro biseriali circiter* 10-*phyllo.*

2. W. *biflora* (DC. prod.)——Frequens in provincia Tanjore versus oppidum Negapatam.—*Wight. cat. n.* 1448.—Verbesina biflora, *Linn. sp.* 1272; *Linn. herb.* (ex *Wight.*); *Wall.! cat. n.* 3207 (excl. *b*).—Acmella biflora, *Spreng. syst.* 3. *p.* 591.—*Rheede, Mal.* 10. *t.* 40.

Verbes. biflora, Roxb. fl. Ind. 3. p. 440, also belongs to this species.—We scarcely know why De Candolle excludes Wall. L. n. 3207. *b*, and refers it to the next species: part of the specimens Dr Wight sent to Dr Wallich marked "*Verb. biflora*," and therefore forming the letter *b* of Dr Wallich, have been recognised by De Candolle himself as his *Wedelia biflora*; they are n. 1448 of Wight's catalogue. Dr Wight sent at the same time specimens of another plant marked "*Verbesina*" simply; this is *W. urticæfolia* β; it is possible that Dr Wallich may have mixed the two, and sent only the latter to De Candolle.

3. W. *urticæfolia* (DC. prod.:) herbacea subscandens, foliis petiolatis ovato-lanceolatis grosse et subinæqualiter serratis utrinque strigoso-hispidis basi breve apice longe acuminatis et sæpe incurvis, pedunculis solitariis 1-cephalis, invol. squamis 10 biseriatis acuminatis dorso pubero-strigosis, paleis acuminatissimis, achæniis exaristatis.

α, *Wallichii.*—Wedelia Wallichii, *Less. in Linnæa*, 1831, *p.* 162.

β, *Wightii*, pedunculis folii circiter longitudine, foliis basi breviter acuminatis parcè strigosis.——In collibus Dindigul.—*Wight. cat. n.* 1449.—Verbesina urticæfolia, *Wight! herb. olim.*—Verbesina biflora, *Wall.! cat. n.* 3207, *b*, et forte *d*.

γ? *scaberrima.*—Verbesina scaberrima, *Wall.! cat. n.* 3201.—Verbesina urticæfolia, *Blume! bijdr.* 911.—An species propria?

III. RAMTILLA. *DC. diss. soc. gener.* 1833, *cum icon.*

Capitulum multiflorum heterogamum, fl. radii ample ligulatis fœmineis sterilibus, disci hermaphroditis tubulosis 5-dentatis. Recept. paleis membranaceis obtusis 3-lineatis inter flores onustum. Invol. squamis 5 ovatis foliaceis subuniseriatis disco æqualibus. Cor. omnes ima basi torquari pilorum articulatorum confertorum manicatæ, tubulosæ ad faucem abrupte dilatatæ et pilos paucos conformes extus gerentes. Stylus radii tenuissimus glaber interdum nullus, disci stigmata apice appendiculata hispidissima exserta gerens. Achænia sessilia oblonga glabra calva.—Herba indica trichotoma erecta. Folia opposita subcordato amplexicaulia lanceolata remote serrata scabrida. Pedunculi 1-cephali sæpe terni. Flores lutei. Achænia intus oleosa.

1. R. *oleifera* (DC. l. c.).—*Wight. cat. n.* 1450.

α, *sativa;* foliis elongato-lanceolatis grosse serratis, ramulis floridis elongatis subcorymbosis.——Culta ad eliciendum oleum in agro Mysorensi notum sub nom. *Ram-Til* seu *Ram-Tilla*, et in Bengalo *Kuts-yelloo, Werrinua*, etc.—Verbesina sativa, *Roxb. cat. calc.* 62; (*fl. Ind.* 3. *p.* 441; *in E. I. C. mus. tab.* 1644; *Wall. L. n.* 3206); *Sim's bot. mag. t.* 1017.—Bidens? Ramtilla, *Wall.! cat. n.* 3191.

This is *Verb. sativa*, Roxb. fl. Ind. 3. p. 441: in E. I. C. mus. tab. 1644; Wall. L. n. 3206.

β, *angustior;* foliis lineari-lanceolatis subdentatis, ramulis axillaribus floriferis brevissimis.—Helianthus oleifer, *Wall.! cat. n.* 3194.

IV. BIDENS. *Linn. gen. n.* 932; *Less. syn.* 230.—Bidens *et* Kerneria, *Cass. dict.*

1. B. *Chinensis* (Willd. sp. 3. p. 1719.)——Ad montes Dindygul, alt. 3500 –4000 ped.—*Wight. cat. n.* 1451.—B. Chinensis, *Wall.! cat. n.* 3189. *d.*

B. bipinnata, Roxb. fl. Ind. 3. p. 411 (in E. I. C. mus. tab. 1635), and Pluk. t. 22. f. 4, seem to be this plant.

2. B. *Berteriana* (Spreng.! syst. 3. p. 454.)——In Porto-Rico (Bert.!), insula Trinitatis (Sieb.! n. 162), in Martinica (forte semina introducta ex Duby! herb.), in India Mader. ex seminibus a provinc. Barrahmal allatis (Wight! herb.).—*Wight. cat. n.* 1452.

3. B. *grandiflora* (Balb.)——In Mexico spontanea, in Ind. or. culta.—*Wight. cat. n.* 1453.

V. GLOSSOGYNE. *Cass. dict.* 58. *p.* 476.

1. G. *pinnatifida* (DC. prod.:) achæniis aristis valde divaricatis (nec ut in *G. tenuifolia* erectis).——Ad Samulcottah, Chingleput prope Madras, &c.—*Wight. cat. n.* 1454.—Bidens pinnatifida, *Wall.! cat. n.* 3190.—Corcopis? Baccana, *Heyn.! herb.*

This is *Zinnia bidens*, Retz, according to Roxb. fl. Ind. 3. p. 435; in E. I. C. mus. tab. 436.—"Willdenow has by some unaccountable mistake quoted Retzius's *Z. bidens*, which I know to be this plant, as I was with Koenig when he first discovered and described it for *Verbesina Boswallia*, a very ramous, cæspitose plant; whereas this is rigidly erect."—ROXB.

VI. XIMENESIA. *Cav. ic.* 2. *p.* 60. *t.* 178.

1. X. *encelioides* (Cav. l. c.)——In India or., sed verisim. introducta. Culta enim adest in horto Benghalensi et certe spontanea in Mexico.

VII. SPILANTHES. *Jacq. amer.* 212.—Spilanthes *et* Acmella, *Rich.; Pers.; Cass.*

1. S. *calva* (DC. prod.:) caule adscendente basi repente apice hirsutulo, foliis petiolatis ovatis obtusis acutisve serratis crenatisve glabriusculis basi ciliatis, pedunculis folio triplo longioribus, capitulis ovatis discoideis, achæniis glabris calvis.——Prope Cunnawady et in agro Mysorense.—*Wight. cat. n.* 1456.—Spilanthes acmella, *Wall.! cat. n.* 3285. *f, h.*—Forte nimis *Sp. pseudo-acmellæ* affinis?

2. S. *oleracea* (Linn.)—Verisim. culta.—*Wight. cat. n.* 1457.

This is also *S. oleracea*, Roxb. fl. Ind. 3. p. 410.

VIII. GLOSSOCARDIA. *Cass. dict.* 19. *p.* 62.

1. G. *Bosvallia* (DC. prod.)——In dumetis et aridis circa Madras.—*Wight. cat. n.* 1458.—Vesbesina Bosvallia, *Linn. f. suppl.* (1781) *p.* 379.—Zinnia bidens, *Retz. obs.* 5. (1789) *p.* 28 (ex *Less.*).—Verbesina Boswallia, *Willd. sp.* 3. *p.* 2223; *Linn. herb.* (ex *Wight!*).—Pectis meifolia, *Wall.! cat. n.* 3198.—Glossocardia linearifolia, *Cass. dict.* 19. *p.* 62 (sed folia minime linearia, potius partita in lobos lineares).

To this belongs *Verbesina Boswallia*, Roxb. fl. Ind. 3. p. 443.

IX. CHRYSANTHEMUM. *Less. syn.* 253.—Chrysanthemi sp. *Linn.*

1. C. *Indicum* (Linn. sp. 1253; Sab. in trans. hortic. soc. 4. p. 326. t. 12.

et 13; soc. Linn. 14. p. 144. *non* Sims. *nec* Thunb.)——In montibus Dindigul, an vere indigenum?—*Wight. cat. n.* 1459.

C. Indicum, Wall. L. n. 3227; Roxb. fl. Ind. 3. p. 436, and Rheed. Mal. 10. t. 44, belong here.

X. ARTEMISIA. *Linn. gen.*

1. A *parviflora* (Don. prod. fl. Nep. 181.)——In India or.—*Wight. cat. n.* 1460.

2. A. *glabrata* (Wall.! cat. n. 3303.)——In India or.—*Wight. cat. n.* 1461. —A. parviflora, *Roxb.?* ex *Wall.!* et *Wight.*

Dr Wallich appears merely to doubt if his n. 3303. *a.* be not referable to *A. parviflora,* Wall. L. n. 3298 (Roxb. fl. Ind. 3. p. 420, and in E. I. C. mus. tab. 1638.)

3. A. *Indica* (Willd. sp. 3. p. 1846; Besser. tent. abr. p. 55. n. 34. descr.) ——An var. Americanæ certe conspecificæ?—In Mysore (Heyne).—*Wight. cat. n.* 1462.—*Wall.! cat. n.* 3293.

This is also *A. Indica,* Roxb. fl. Ind. 3. p. 449; in E. I. C. mus. tab. 1169; Rheede's Mal. 10. t. 45, and perhaps Pluk. t. 15. f. 1.—We cannot distinguish *A. grata,* Wall. L. n. 3294.

4. A. *pallens* (Wall.! cat. n. 3302; Bess. tent. abr. p. 33. n. 74. descr.)

XI. MYRIOGYNE. *Less. in Linnæa* 1831, *p.* 219. *syn.* 266.—Cotulæ sp. *auct.*

1. M. *minuta* (Less. l. c.)——In humidis Indiæ or.—*Wight. n.* 1464.—Cotula minuta, *Forst. prod.* 301.—Grangea minuta *et* cuneifolia, *Poir. dict. suppl.* 2. *p.* 825.—Cotula minuta *et* cuneifolia, *Willd. sp.* 3. *p.* 2189.—Artemisia minima, *Thunb. fl. Jap.* 311 (non *Linn.*)—Cotula sternutatoria, *Wall.! cat. n.* 3239.—Variat glabra aut lanuginosa.

To this belongs *Artemesia sternutatoria,* Roxb. fl. Ind. 3. p. 423; in E. I. C. mus. tab. 1286.

XII. HELICHRYSUM. *Less. syn. p.* 274.

N. B. *Huc relatæ sunt juxta cl. Lessing species capitulis aut homogamis aut series fl. fœm.* 1–2 *tantum gerentibus, sed genus ulteriori examine eget quoad caract. genericum et sectiones.*

1. H. *buddleioides* (DC. prod.:) caule suffruticoso erecto ramoso apice præsertim lanuginoso, foliis sessilibus ovali-lanceolatis acuminatis 7–9-nerviis supra glabris subtus cano-tomentosis integerrimis, corymbo composito polycephalo ad apices caulis et ramorum, capitulis ovatis dense confertis homogamis, invol. squamis ovalibus obtusis subæqualibus disco paulo longioribus.——In Neelgherry (Wight! herb.)—*Wight. cat. n.* 1465.—Gnaphalii sp. *Wall.! cat. n.* 2932 (ex quo vix Indica).—Invol. et flores flavi. Recept. alveolatum breviter fimbrilliferum. Paulo accedit ad *Hel. nudifolium, plantaginifolium,* inter *Lepiclines* admissum.

2. H. *Notonianum* (DC. prod.:) totum lanatum, caule suffruticoso, ramis usque ad apicem foliosis, foliis lato-linearibus oblongisve obtusis margine revolutis lanugine crassissimis sessilibus subdecurrentibus, corymbo terminali composito denso, capitulis heterogamis, inv. ovati squamis multiserialibus imbricatis albo-scariosis acutis apice subcrispis demum stellato-patulis.——In Neelgherry.—*Wight. cat. n.* 1467.—Gnaphalium Notonianum, *Wall.! cat. n.* 2952, et alterum innominatum, *n.* 2933.—Rec. nudum vix. alveolatum. Invol. albidum. fl. fœm. 1-seriales.

3. H. *leptophyllum,* totum sericeo-canum, caule tereti ramoso erecto, foliis distantibus sessilibus linearibus acuto-mucronatis integerrimis patulis, corym-

bo terminali composito subumbellato, capitulis pedicellatis heterogamis, invol. squamis oblongo-linearibus subacutis albo-coloratis squarroso-reflexis. ——In Neelgherry.—*Wight. cat. n.* 1467.—Rec. nudum punctulatum. Invol. album. Fl. fœm. 1–2 serialis.

XIII. GNAPHALIUM. *Less. syn.* 328.

* *Suffrutescentia (saltem basi), Helichrysoidea.*

1. G. *hypoleucum* (DC. prod. :) caule erecto tereti inferne scabrido superne ramoso-tomentoso, foliis linearibus acuminatis margine subrevolutis supra scabridis subtus niveo-tomentosis nervo medio scabrido basi adnata semiamplexicaulibus subdecurrentibus, capitulis ad apices ramorum congestis subsessilibus, glomerulis corymboso-paniculatis, invol. squamis flavo-scariosis ovali-oblongis obtusiusculis disco paulo longioribus.——In Neelgherry.—*Wight. cat. n.* 1468.—Species elegans foliis subtus albis et invol. flavis facile distincta. Fl. fœm. pluriseriales.

2. G. *subdecurrens* (DC. prod. :) totum niveo-lanatum, caule subherbaceo, foliis infer. oblongo-subovatis super. oblongo-linearibus acutis basi semiamplexicaulibus vix subdecurrentibus, capitulis ad apicem in corymbum compositum dense confertis, invol. ovati squamis subcariosis albido-coloratis ovati-oblongis acutis disco longioribus.——In montibus Dindygul.—*Wight. cat. n.* 1469.—Gnaphalium semidecurrens, *Wall.! cat. n.* 2947, *b*.—Habitus Antennariæ decurrentis, *DC.* (*Wall. cat.* 2947, *a*), sed caracter omnino Gnaphalii. Invol. in sicco albido-stramineum.

3. G. *aristatum* (DC. prod. :) caule basi suffruticuloso ramoso erecto ad apicem usque folioso basi scabro apice tomentoso, foliis linearibus elongatis aristato-mucronatis marginibus revolutis basi dilatato cordatis breviter decurrentibus supra scabris nervo medio subtomentoso subtus cano-tomentosis nervo medio scabro, capitulis in corymbum terminalem compositum densissimum confertis, invol. squamis ovalibus obtusis albo et roseo coloratis glabris. ——In Neelgherry.—*Wight. cat. n.* 1470.—Fl. in invol. 15–18 quorum 5–6 hermaph., cæteri fœm. 2–3 seriales.

4. G. *Wightianum* (Wall.! cat. n. 2940 :) caule basi suffruticoso erecto ad apicem folioso basi piloso-scabro apice lanuginoso, foliis sessilibus oblongo-linearibus obtusis supra piloso-scabris subtus cano-lanuginosis nervo sæpe subtus prominulo scabro summis apice calloso-subuncinatis, corymbo terminali vere composito sed densissime polycephalo, squamis involucri oblongis acutiusculis albo-coloratis disco paulo longioribus.——In Neelgherry.—*Wight. cat. n.* 1471.—Fl. flavi, ext. fœminei pluriseriales, centr. hermaphr. 10–12.

5. G.? *ellipticum* (DC. prod. :) totum niveo-tomentosum, caule basi suffruticuloso ramoso humili, foliis adnatis seu breviter decurrentibus ellipticis mucronulatis integerrimis tomento-absterso 5–7-nerviis, capitulum in corymbum ovatum terminalem foliis cinctum densissimum confertis, invol. squamis ovalibus acutis apice scariosis ex rufo albidis.——In Neelgherry.—*Wight. cat. n.* 1472.—Fl. ex statu juniori videntur exteriores fœminei 2–4 seriales.

6. G. *Neelgherryanum*, caule fruticoso multicipiti humili, ramis floridis erectis tomentosis, foliis infer. densissime confertis retrorsum imbricatis linearibus subobtusis infimis senescentibus glabratis super. secus ramos floridos erectis tomentosis acutis subdistantibus, capitulis in corymbum terminalem densum confertis, invol. squamis oblongo-linearibus subacutis disco longioribus albido coloratis.——In Neelgherry ad altitud. 7000 ped. et ultra.—*Wight. cat. n.* 1473.—Cæspes durus densus ramosissimus canus. Fl. fœm. 2–3-seriales. Pappi setæ apice barbellato penicillatæ. Habitus Helichrys. frigidi.

* * *Herbacea, Filaginoidea.*

7. G. *pallidum* (Wall.? cat. n. 2953, excl. litt. *a*:) **lanatum, caulibus pluribus herbaceis adscendentibus erectis simplicibus, foliis infer. lanceolatis subacutis planis caulinis late sessilibus summis linearibus, ramulis floridis apice subumbellatis, capitulis in glomerulos ad axillas foliorum et apices ramulorum subaggregatis, invol. squamis oblongis ellipticis obtusis ima basi lanatis cæterum glaberrimis stramineo-rufis.**——Ad montes Circars; *Heyne* ex *herb. Rottl.*—*Wight. cat. n.* 1474.—Affinis G. luteo-albo sed, invol. squamæ obtusæ, et rufescentes, glomeruli subumbellati, etc.

To this belongs *G. Orixense*, Roxb. fl. Ind. 3. p. 425; in E. I. C. mus. tab. 426.

8. G. *indicum* (Lin. sp. 1200.)——In locis humidis Ind. or. frequentissimum.—*Wight. cat. n.* 1475; *Wall.! cat. n.* 2954.

G. strictum, Roxb. fl. Ind. 3. p. 424, in E. I. C. mus. tab. 425, *G, multicaule*, Willd. sp. 3. p. 1888, and Pluk. t. 187. f. 5 belong to this species.

XIV. FILAGO. *Tourn. inst. t.* 259; *Lin. sp.* 1311 (excl. sp. 1. et 7); *Gærtn. fr.* 2. *t.* 166 (non *Cass.* nec *Less.*)—Gifola, Oglifa *et* Logfia, *Cass.*

1. F. *prostrata* (DC. prod.:) **multicaulis diffuso-prostrata tota lanuginosa ramosissima, foliis ellipticis mucronatis in petiolum angustatis, capitulis dense corymboso-aggregatis foliis floralibus stipatis, invol. squamis mucronatis.**——In humidis prov. Tanjore prope Negapatam frequens. Circars; *Roxburgh*; *Heyne.*—*Wight. cat. n.* 1476.—Gnaph. prostratum, *Roxb.! herb.*; *Wall.! cat. n.* 2955.—Habitus Gn. crispatuli, sed caract. florales fere Filaginis arvensis.

The name *G. prostratum*, although originally applied to this species by Roxburgh, both in his own herbarium and in the India House museum, tab. 427, was afterwards altered in the Hort. Benghalensis, and in the Flora Indica (3. p. 425), to that of *G. depressum.* It is therefore to be regretted that Wallich and De Candolle have preferred the manuscript to the published one.

XV. CARPESIUM. *Linn. act. Ups.* 1741, *p.* 81; *gen.* 948.

1. C. *Nepalense* (Less. in Linnæa 1831, p. 234.)——In Nepalia; *Wall.!*—*Wight. cat. n.* 1477.

XVI. SENECIO. *Lin. gen. n.* 955; *Less. syn. p.* 391.

S. flexuosus, Wight, cat. n. 1506 (*Cacalia antieuphorbium*, herb. Rottl., and *C. Kleinia*, Linn., according to Klein's herb.), is here omitted: it is certainly a native of the Mauritius, and not of India.

* *Capitulis discoideis.*

1. S. *corymbosus* (Wall.! cat. n. 3121:) **caule scandente tereti subaraneoso, foliis petiolatis exstipulatis cordatis suborbiculatis breviter acuminatis subserratis supra glabriusculis subtus dense tomentosis basi 5–7-nerviis, corymbis axillaribus et terminalibus densis polycephalis, invol. 8-phyllo basi bracteolato, ligulis nullis, achæniis glabris.**——Ad Neelgherry.—*Wight. cat. n.* 1478. Flor. tubulosi 10.

* * *Capitulis radiatis.*

2. S. *candicans* (Wall.! cat. 3123:) **totus tomentoso-canus sarmentosus, ramis striatis, foliis petiolatis stipula reniformi utrinque auriculatis cordatis acutis serratis supra arachnoideis subtus niveis, panicula corymbosa, bracteis lineari-subulatis, pedicellis divergentibus, invol. cano-campanulato basi parce bracteolato, ligulis 6 oblongis planis, achæniis glabris.**——In Neelgherry.—*Wight. cat. n.* 1479.—Cineraria sarmentosa, *Lesch.! herb.*, ex quo vulgo dicitur Kouri-gueda.

3. S. *Wightianus* (DC. prod.:) **glaber, ramis scandentibus angulato-striatis,**

foliis petiolatis ovato-elliptico-ve lanceolatis acuminatis serratis limbo basi obtuso aut brevissime cuneato, petiolo basi parve auriculato, panicula divaricata, pedicellis apice bracteolatis, capitulis parvis 8–10-floris, ligulis 3–4 parvis. —— In Ceylon, et apud Payanee.—*Wight. cat. n.* 1480.——Cacalia Wightiana, *Wall.! cat. n.* 3165.—Vix S. scandens, *Don, prod. fl. Nep. p.* 178? (non *Wall.*)

4. S. *lavandulæfolius* (Wall.! cat. n. 3130:) caule adscendente tereti striato hirsuto, foliis confertis oblongo-linearibus integerrimis margine revolutis subtus tomentosis supra hirsutis aut hispidis, summis linearibus distantibus, racemo corymboso simplici, pedunculis bracteolatis, invol. glabriusculo 15-phyllo calyculato, flosculis circiter 40, ligulis 15 patulis longis 4-nerviis. ——In Neelgherry.—*Wight. cat. n.* 1481.

β, *Wightii;* foliis utrinque cano-tomentosis, racemo oligocephalo capitulis breve pedicellatis.

We consider the variety β to be an accidentally starved state of the other, and not a distinct variety.

5. S. *Neelgherryanus* (DC. prod.:) caule basi suffrutescente erecto scabrido striato, foliis lineari-lanceolatis acutis subtus hirsuto-tomentosis supra scabris, imfimis basi attenuatis semipinnatifidis mediis sessilibus dentatis summis auriculato-semiamplexicaulibus subintegerrimis, corymbo oligocephalo, pedicellis longis apice bracteolatis, invol. squamis linearibus apice vix acutis ciliatis, ligulis 12–14 planis.——In Neelgherry.—*Wight. cat. n.* 1482.—Affinis S. saxatili, *Wall.! cat. n.* 3131.

XVII. DORONICUM. *Linn.; Less. syn. p.* 390.

6. D. *Arnottii* (DC. prod.:) caule erecto tereti striato subsimplici inferne hirto apice glabriusculo, foliis ad basin confertioribus cordato-semiamplexicaulibus oblongis 3–5 nerviis obtusis dentatis utrinque hirtis, corymbo oligocephalo terminali, bracteolis lineari-subulatis, invol. subcalyculati squamis linearibus circ. 15, ligulis 10 ovalibus 6-nerviis forte stylo abortivo sterilibus. ——In Neelgherry et Dindygul.—*Wight. cat. n.* 1483.—Stigmata inclusa. Pappus rufescens. Flosculi innumeri.

7. D. *Wightii* (DC. prod.:) glaber, caule adscendenti-erecto basi striato-angulato subsimplici, foliis lanceolatis acutis grosse dentatis margini subrevolutis infimis basi attenuatis super. semiamplexicaulibus, corymbo oligocephalo, pedicellis apice bracteolatis, invol. squamis linearibus subacutis, ligulis 8–10 planis 5–7-nervis.——Ad Neelgherry.—*Wight, cat. n* 1484.

β, *laxus;* corymbo ramoso magis divaricato, capitulis numerosioribus minoribus.—*Wight. cat. n.* 1484, *b.*

XVIII. NOTONIA. *DC. prod. vol.* 5.

This genus has been published by De Candolle in Guillemin's Archives de Botanique for December 1833. We had previously made use of the same name (in Wight's Lithographic Catalogue, Nos. 870, 872) for plants belonging to the Leguminosæ: the description of our genus was also printed (although the work was not published for some months afterwards), in our Prodromus Fl. Pen. Ind. Orientalis. We have, however, given up any claim in favour of our genus, and applied to it the name of Johnia (see Prod. Fl. Pen. Ind. Or. 1. p. 449).

Capitulum multiflorum homogamum, fl. tubulosis 5-dentatis. Invol. cylindricum 1-seriale ecalyculatum, bracteolis nullis aut secus pedicellum paucissimis. Recept. planum alveolare, alv. marginibus productis fimbrilliferum. Styli rami in conum brevem hispidum apice (ut in Emilia) productum. Achænia teretia multistriata glabra. Pappus multiserialis, setis filiformibus hinc inde barbellatis.—Suffrutices indici carnosi glabri facie Cacaliæ Kleiniæ,

Folia oblonga aut obovata plana subcarnosa, integerrima. Corymbi pauciflori. Capitula inter genera affinia magna. Genus dicatum cl. Notoni qui Indiæ stirpes strenue collegit et speciem generis typum cum cl. Wight observavit.

1. N. *grandiflora* (DC. prod.:) glaberrima fruticoso-carnosa, ramis teretibus foliis oblongis obovatisve integerrimis, corymbo oligocephalo, involucro floribus breviore.——In Neelgherry, ad montes Dindygul, apud Pathoocottah, et in provincia Mysore, in dumetis.—*Wight. cat. n.* 1485.—Cacalia grandiflora, *Wall.! cat. n.* 3147.—Species polymorpha forte in posterum dividenda. Cor. ex sicco videntur flavæ, cum ovario fere pollicem longæ.

NB. Altera generis species (*N. crassissima*) est *Cacalia crassissima*, Wall.! cat. n. 3155.

XIX. EMILIA. *Cass. dict.* 14. *p.* 405. *v.* 34. *p.* 393, *et* 48. *p.* 449, 461.

1. E. *sonchifolia* (DC. prod.)——In Ind. or. frequens.—*Wight. cat. n.* 1486. —Cacalia sonchifolia, *Lin. sp.* 1169.—Cacalia sonchifolia et C. glabra, *Wall.! cat. n.* 3144 (excl. litt. *d*), et *n.* 3145.—Cor. purpureæ aut aurantio-purpureæ in invol. 30–50.

This is *Cacalia sonchifolia*, Roxb. fl. Ind. 3. p. 413, in E. I. C. mus. tab. 422, Rheed's Mal. 10. t. 68, and Pluk. t. 144. f. 4.

2. E. *scabra* (DC. prod.:) caule adscendente basi dense hirtello folioso apice nudo lævi, foliis infimis . . . . . caulinis cordato-semiamplexicaulibus ovali-lanceolatis subobtusis dentatis utrinque pube sparsa scabris, corymbo terminali 5–7-cephalo, invol. flosculis vix breviore, flosculis in invol. circiter 100.——In Ind. or.—*Wight. cat. n.* 1487.

XX. GYNURA. *Cass. dict.* 34. *p.* 391 ; 48. *p.* 448. *et* 458.

1. G. *nitida* (DC. prod.:) glabra, caule basi crassiusculo tereti, ramis elongatis subangulatis apice subnudis, foliis lanceolatis utrinque attenuatis grosse serratis, corymbo terminali 5–7-cephalo, invol. cylindrico floribus paulo breviore bracteolis subulatis quadruplo longiore.——Ad montes Dindigul, altit. 4000 ped.—*Wight. cat. n.* 1488.—Cacalia incana, *Heyn.!; Wall.! cat. n.* 3158 (an Lin. sp. 1169, sed quid in planta incani?). Stigma generis sed brevius quam in cæteris speciebus.

2. G. *lycopersicifolia* (DC. prod.:) caule herbaceo erecto ramoso-striato glabriusculo, foliis pinnatilobatis basi ample et rotunde auriculatis utrinque pube brevissima velutinis, lobis lanceolatis hinc inde dentatis, capitulis ad apices ramorum 3–5 corymbosis pedicellatis, invol. cylindrico-glabro flosculis subæquali bracteolis subulatis triplo longiore.——Ad Courtallum et in montibus Dindigul alt. 2000 ped.—*Wight. cat. n.* 1489.—Cacalia laciniata, *Wall.! cat. n.* 3153 (non *Jacq.*)—Cacalia pinnatifida, *Wight! herb. olim* (non *Lour.* nec *Lin. f.*)—Cacalia lycopersicum, *Heyn.! herb.*—Var. rarius auriculis parvis acutis.

TRIB. V.—CYNAREÆ. *Less. syn. p.* 4.

I. ECHINOPS. *Lin. gen. n.* 999.

1. E. *echinatus* (Roxb.! herb., Wall.! cat. n. 2985:) foliis pinnatifidis supra viscoso-pubescentibus subtus cano-tomentosis laciniis ovato-lanceolatis sinuatis planis, glomeruli invol. generali squamis brevibus 3-cuspidatis spinosis, capituli invol. partiali in iisdem glomerulis diverso, nunc setis externis longissimis intimis concretis fructu villosissimo, nunc setis ext. brevioribus folio int. uno in cornu longe exserto fructu glaberrimo (abortivo?)——In

pascuis aridis agri Mysorensis atque Hydrabadensis, prope Baitmungolany.—*Wight. cat. n.* 1490.

This is *E. echinatus*, Roxb. fl. Ind. 3. p. 4447; in E. I. C. mus. tab. 1645.

II. CENTAUREA. *Less. syn. p.* 7.—Centaurea sp. *Linn.*

1. C. *pulchra* (DC. prod.:) caule erecto ramoso foliisque uniformiter albido-tomentosis, foliis lato-linearibus basi attenuatis apice acuminatis integerrimis aut hinc inde rarissime denticulatis, fructus pappo duplici, ext. paleis linearibus vix acutis regulariter imbricatis et successivi longioribus achenii longitudinem æquantibus, interiore paucisecto.——Culta in hortis Indicis.—*Wight. cat. n.* 1491.—C. pulcherrima, *Wight! herb. olim* (non *Willd.*)—C. cyanus, *Wall.! cat. n.* 2981, *b.*—Nimis affinis C. cyano et forte varietas? Pappus in fructu equidem immatura descr. videtur diversus. Capitulum paulo majus. Squamæ invol. margine albæ aut rufæ, nec nigræ.

III. CARTHAMUS. *Tourn. inst. t.* 258.—Carthami sp. *Linn.*

1. C. *tinctorius* (Lin. sp. 1192).—Cultus in prov. australibus Indiæ or. ad oleum e seminibus eliciendum.—*Wight. cat. n.* 1492.

To this belongs *C. tinctorius*, Roxb. fl. Ind. 3. p. 409; Wall. L. n. 2959.

IV. CIRSIUM. *Tourn. inst. t.* 255: *DC. diss. comp.* 1. *p.* 22 (non *Less.*)—Cnicus, *Schreb. gen. n.* 1255 (non *Vaill.*)

1. C. *argyracanthum* (DC. prod:) foliis semiamplexicaulibus sinuato-pinnatifidis ciliato-spinulosis lobis in spinas validas productis subtus cauleque arachnoideo-villosis, capitulis paniculato-congestis, bracteis multifidis spinosissimis, invol. squamis validè spinosis.——In Neelgherry.—*Wight. cat. n.* 1493.—Carduus argyracanthus, *Wall.! cat. n.* 2903.—Cor. ex sicco purpureæ. Bractearum pars foliacea brevissima, spinæ elongatæ validissimæ. Ad Cirsii sectionem Eriolepis referendum juxta C. strigosum, *Bieb.*

V. TRICHOLEPIS. *DC. prod.*

Capitulum multiflorum homogamum. Invol. hemisphærici squamæ imbricatæ numerosissimæ lineari-setaceæ apice criniformes subrecurvæ inermes minime uncinatæ. Recept. fimbrillæ in setas tenuissimas fissæ. Cor. 5-fidæ subregulares. Stam. filamenta papilloso-pilosiuscula. Antheræ append. acutis, caudis parvis sublaceris. Achænium oblongum hilo laterali apice margine circulari coronatum. Pappus varius.—Herbæ indicæ glabriusculæ inermes. Folia sessilia indivisa lanceolata aut oblonga sæpe punctata. Capitula ebracteata. Cor. ex sicco ochroleucæ aut pallide rubentes.—Præter duas sequentes huc referendi *Carduus trichocephalus*, Wall., et *Carduus elongatus*.

* *Pappo paleaceo, paleis ciliolatis.*

1. T. *radicans*, caule ramoso basi radicante angulato, foliis oblongo-linearibus argute et grosse aristato-serratis glabris punctatis, invol. ovati squamis araneosis appendice subulata glabra.——In India Maderaspatana.—*Wight. cat. n.* 1494.—Carduus radicans, *Roxb.* ex *Wall. cat. n.* 2907.

This is *Carduus radicans*, Roxb. fl. Ind. 3. p. 408; in E. I. C. mus. tab. 1634.

** *Pappo nullo.*

2. T. *angustifolia* (DC. prod.:) glabra, caule ramoso angulato, foliis elongato-linearibus acuminatis punctatis rariter aristato-serratis, capitulis ovatis glabris, achænio epapposo vix apice marginato.——In montibus Travanco-

rensibus.—*Wight. cat. n.* 1496.—Carduus indicus, *Klein! in herb. Wight.*—C. nitidus, *Wall.! cat. n.* 2908.—Serratula indica, *Willd. sp.* 3. *p.* 1642.

TRIB. VI.—MUTISIACEÆ. *Less. syn. p.* 92.

I. DICOMA. *Cass. dict.* 13. *p.* 194; 47. *p.* 503; *Less. syn.* 109.

1. D. *lanuginosa* (DC. prod.:) involucri ovato-oblongi squamis extus glabriusculis, pappi paleis serratis fructu villosissimo vix duplo longioribus.—In pascuis et apricis littoris Coromandeliani.—*Wight. cat. n.* 1497.—Xeropappus lanuginosus, *Wall.! cat. n.* 2980.—Acelepis lanata, *Spreng. in litt. ad Wall.*—Onopordon lanatum, *herb. Madr.* (ex *Wall.*)—Valde accedit ad D. tomentosam, *Cass.*; sed caract. indicatis facile differt.

TRIB. VII.—CICHORACEÆ. *Juss. gen.* 168; *Less. syn.* 126.—Lactuceæ, *Cass.*

I. CICHORIUM. *Linn. gen. n.* 921.

1. C. *Cosnia* (Ham.! in Wall.! cat. n. 3241:) caule herbaceo erecto lævi hinc inde piloso, foliis oblongis sinuatis aut subpinnatifidis scabridis floralibus basi auriculatis, capitulis axillaribus pluribus congestis sessilibus.—In variis Indiæ locis a Wallichio indicata, sed forte vix indigena ex Wight herb. —*Wight. cat. n.* 1498.—C. endiviæ affine.

II. PICRIS. *Linn. gen. n.* 907.

1. P. *hieracioides* (Linn. sp. p. 1114).

β, *indica;* corymbo magis divaricato.—Ad Courtallum.—*Wight. cat. n.* 1499.—Forte P. hamulosa, *Wall.! cat. n.* 3255.—In Nepaulia locta ab hac non differt?

III. LACTUCA. *Linn. gen. n.* 909; *Less. syn.* 135.

* *Achænii collo elongato.*

1. L. *sativa* (Linn. sp. 1118).—In Ind. or. culta.—*Wall. cat. n.* 3244.

This is also *L. sativa*, Roxb. fl. Ind. 3. p. 403.

** *Achænii collo brevi.* Mycelis, *Cass. dict.*

2. L. *glabra* (DC. prod.:) glaberrima, caule nudo parce ramoso, foliis paulo longiore, foliis elongato-linearibus integerrimis acutis, capitulis longe pedicellatis cylindraceis subcorymbosis.—In Ind. or.—*Wight. cat. n.* 1500.

3. L. *remotiflora* (DC. prod.:) caule herbaceo tereti glabro subnudo parcè ramoso, foliis subradicalibus membranaceis amplexicaulibus, obovatis vix subsinuatis margine setoso-ciliatis cæterum glabris, capitulis secus ramos subsolitariis distantibus pedicellatis cylindraceis.—In Peninsula indica frequens.—*Wight. cat. n.* 1501.

From a specimen similarly named in Dr Wight's herbarium, this appears to be *Cacalia sonchifolia*, Wall. L. n. 3144, *d.*

4. L. *runcinata* (DC. prod.:) caule erecto glabro tereti apice nudo, foliis subradicalibus rigidulis runcinatis grosse setoso-ciliatis cæterum glabris amplexicaulibus, capitulis secus ramos distantibus subgeminis breviter pedicellatis cylindraceis.—Frequens in ruderatis circa Negapatam.—*Wight. cat. n.* 1502.—Prenanthes sonchifolia, *Heyn. in Wall.! cat. n.* 5266; *Willd. sp.* 3. *p.* 1541.—Non est Lactuca sonchifolia auct.

P. racemosa, Roxb. fl. Ind. 3. p. 405, or *Sonchus calyculatus*, Roxb. in E. I. C. mus. tab. 418, belongs to this species.

5. L. *napifolia* (DC. prod.:) glabra caulibus erectis nudis ramosis, foliis radicalibus petiolatis lyrato-pinnatilobatis mucronatis lobis oblongis terminali majore, capitulis longe pediculatis parvis, invol. subuniseriali.——In ruderatis circa Negapatam.—*Wight. cat. n.* 1503.—Prenanthes napifolia, *Wall.! herb.*

6. L. *sarmentosa* (Wight herb.!:) glabra, caule flagellifero procumbente hinc inde radices foliaque agente, foliis confertis sinuato-pinnatifidis lobis rotundatis, pedunculis (scaposis) 1-cephalis apice bracteolas squamosas subimbricatas margine scariosas gerentibus, invol. cylindraceis.——Frequens in arenosis ad littora maris versus Negapatam.—*Wight. cat. n.* 1504. Prenanthes sarmentosa, *Willd. sp.* 3. *p.* 1540; *Wall.! cat. n.* 3262.

The Coromandel specimens mentioned by Roxburgh under *Prenanthes asplenifolia* (fl. Ind. 3. p. 404, or *Hieracium stoloniferum*, Roxb. in E. I. C. mus. tab. 1281), belong to this species.

IV. SONCHUS. *Linn. gen. n.* 908.

1. S. *oleraceus* (Linn. sp. 1115).——Frequens versus Negapatam.—*Wight. cat. n.* 1505; *Wall.! cat. n.* 3252.

This is *S. oleraceus*, Roxb. fl. Ind. 3. p. 402, or *S. Indicus*, Roxb. in E. I. C. mus. tab. 47; and, judging from Roxburgh's descriptions and drawing, his *S. orixensis*, fl. Ind. 3. p. 402 (in E. I. C. mus. tab. 416), seems also to belong to this variable species.

No. II.

ASCLEPIADEÆ INDICÆ.

[The Peninsular genera and species by Dr Wight and Mr Arnott, almost all the others by Dr Wight alone, and marked *R. W.* The materials principally examined were Dr Wallich's undistributed Asclepiadeæ (quoted *Wall. Asclep.*) in the possession of the Linnean Society, Dr Royle's, and Dr Wight's own collection. We need scarcely say, that the best characters for the species as well as for the genera are to be taken from the flower. Lamarck's and Poiret's species are too obscurely described to be recognised; nor are Roth's characters more intelligible: some of the species, as *Periploca reticulata*, perhaps belong to the *Apocyneæ*.—Mr Brown, in his character of the order (Wern. Soc. Mem. 1. p. 19, and Prod. fl. nov. Holl. 1. p. 458), applies the term *corpuscle* to those firm glandular bodies that are inserted into the angles of the stigma. He also, in the same place, mentions that the pollen-masses are fixed to the *five processes* of the stigma; whence it might be inferred that the *processus* was meant to include both the corpuscle and its projecting filaments to which the masses were attached: in his tabular view, however (as well as in his character of *Oxypetalum* and *Xysmalobium*), this does not appear to be the case, for he there restricts that term to the connecting filaments, "corpuscula juxta basin vel utrinque exserentia processum massam pollinis unicam affigentem:" in a later memoir (his observations on their organs of fecundation) he calls them "arms." Dr Wallich, in his Pl. Asiat. rar., while describing some species of Asclepiadeæ, appears to apply in some instances the term *retinaculum* to each corpuscle with its two processes combined, which last he calls the *crures*: but, in other cases, he uses corpuscle in the sense employed by Mr Brown, and states that the pollen-masses are pedicellate, or attached by five threads. Sprengel calls the corpuscle a *glandula*, and its processes *filamenta*; but under *Oxypetalum* (Genera pl. p. 213), he makes these last synonymous with *retinacula*. In the Orchideæ, Richard restricts *retinaculum* to the gland at the base, and calls the connecting stalk by the name *caudicula*: Mr Lindley retains the latter, but designates the former by *glandula*. The appearance of the parts, however, is so different as scarcely to warrant an application of the same terms in both orders, and it thus becomes desirable that more precise ideas be adopted: we may therefore mention, that, in the following memoir, we have employed *processus* for the whole body inserted on the angle of the stigma, as in Mr Brown's character of the order, and that this is composed of the *corpuscle* of Mr Brown, and its processes or arms, which we shall call *appendiculi*, προσθηκαι, or appendages of the corpuscle.—R. W.—G. A. W.-A.]

I. ASCLEPIADEÆ VERÆ. *R. Brown.*

Massæ pollinis 10, læves, per paria (diversis antheris pertinentia) mediante appendiculo affixæ stigmatis corpusculis 5 sulco longitudinali bipartibilibus. Filamenta connata, extus sæpius corona instructa.

I. CEROPEGIA. *Linn.*—Isaura, *Comm.*—Stephanotis, *Pet. Thouars.*

Calyx 5-partitus. Corolla e basi magis minusve ventricosa infundibuliformis; limbi laciniis angustis apice cohærentibus. Gynostegium inclusum. Corona staminea gamophylla, 5–10–15-lobata, serie simplici vel duplici; lobis ante stamina positis ligulatis. Antheræ apice simplices. Massæ pollinis erectæ, basi affixæ, marginibus simplicibus. Stigma muticum. Folliculi cylindracei, læves. Semina comosa.—Herbæ vel suffrutices, sæpius volubiles, nunc lactescentes.

The staminal crown in this and in most of the other genera of the order consists of a single verticil of 5 leaves opposite the stamens and alternate with the segments of the corolla. In *Ceropegia* the leaflets are usually 3-lobed: the lobes are sometimes distinct to near the base, forming a 15-parted crown; more rarely nearly or altogether united, forming one that is 10- or 15-lobed: and sometimes the lateral lobes become united to the middle one, and thus appear wanting, presenting a 5-lobed crown. When the leaflets are distinct to near the base, the lateral segments approach and partially adhere to the back of the middle or primary one, when we have them apparently in a double series; but when they are united to near the apex, the lateral segments remain distinct, and a single series only is formed.

§ 1. *Corona serie simplici* 10-*loba; lobis alternis brevioribus, ovatis, obtusis, integris vel semibifidis* (*e lobis duobus ad foliola diversa pertinentibus*).

1. C. *lucida* (Wall.:) glabra volubilis, radice fibrosa, foliis e late ovatis in oblongo-lanceolata acuminata, pedunculis multifloris, laciniis calycinis subulatis apice recurvatis, corolla clavata vix basi ventricosa, limbi segmentis attenuatis tubum æquantibus, coronæ stamineæ foliolorum lobis lateralibus supra medium unitis margine pilosis primariis apice recurvis.—*Wall.! pl. as. rar.* 2. *p.* 33. *t.* 139; *Asclep. n.* 2, 9, 14.——Ad montes Silhet et Prome.—(*R. W.*)

2. C. *hirsuta* (W. & A.:) suffruticosa hirsuta volubilis, foliis e cordato-ovatis obtuse acuminatis in anguste lanceolata acuta, pedunculis paucifloris, laciniis calycinis filiformibus, corolla clavata basi ventricosa limbi segmentis sursum latioribus tubo brevioribus, cor. st. foliolorum lobis lateralibus ad medium unitis primariis elongatis apice uncinatis, folliculis rectis glabris.—*Wight! cat. n.* 1510.——Ad Neelgherry.

3. C. *juncea* (Roxb.:) glabra subcarnosa volubilis lactescens, foliis parvis sessilibus lanceolatis acutis, pedunculis paucifloris, laciniis calycinis subulatis, corolla clavata curvata basi ventricosa, limbi segmentis sursum latioribus tubo fere æqualibus pilosis, cor. st. foliolorum lobis lateralibus ad medium unitis primariis apice uncinatis.—*Roxb. Cor.* 1. *p.* 12. *t.* 10; *Willd. sp. pl.* 1. *p.* 1275; *Roem. et Schult. syst. veg.* 6. *p.* 3; *Wall.! Asclep. n.* 5; *Wight! cat. n.* 1511.——Ad montes Circars, in sepibus prope Samulcottah, apud Vellanganny prope Negapatam, Columala, et in provincia Tanjore haud infrequens.

4. C. *Wightii* (Graham:) glabra volubilis, radice tuberosa, foliis ovatis acutis carnosis, corolla basi sphærice ventricoso, tubo gracili, limbi segmentis pubescentibus, cor. st. foliolorum lobis lateralibus ad apicem unitis glabris primariis carnosis lateraliter compressis medio extrorsum curvatis cæteros vix duplo excedentibus.—*Graham! mss.; Wight! in Bot. mag. t.* 3267.

This species is only known from a plant which flowered in the Edinburgh Botanic Garden, and is supposed to have been received from Bombay: it is nearly allied to *C. Africana* of horticulturists (figured in the Botanical Register) in the form of the corolla and crown; that species has also been stated to come from the East Indies: probably both are from the Cape of Good Hope.—(*R. W.*)

§ 2. *Corona staminea serie (ut videtur) duplici; foliolis ad basin fere liberis, lobis lateralibus distinctis primario approximatis ejusque dorso ad basin cohærentibus.*

* *Cor. st. foliolorum lobi laterales ligulati, gynostegio longiores.*

5. C. *elegans* (Wall.:) suffruticosa glabra volubilis, radice fibrosa, foliis oblongis vel oblongo-lanceolatis acuminatis, pedunculis folio brevioribus 1–2-floris (culta 2–6-floris), laciniis calycinis subulatis basi corollæ ventricosa multo brevioribus, cor. tubo clavato, limbi hemisphærici segmentis late ligulatis tubo multo brevioribus longe ciliatis, cor. st. foliolorum lobis lateralibus primariis subduplo brevioribus ligulatis, folliculis gracilibus torulosis curvatis (6–8 uncias longis).—*Wall.! Asclep. n.* 4; *in bot. mag. t.* 3015; *Wight.! cat. n.* 1512.——In montibus Dindygul et Neelgherry.

This very beautiful species is abundant on the Neelgherries, at the height of between 6000 and 7000 feet.

6. C. *macrantha* (Wight:) suffruticosa volubilis, foliis ovatis acuminatis ciliatis, umbellis paucifloris breve-pedunculatis, laciniis calycinis subulatis, corolla basi ventricosa, limbo paullum dilatato segmentis anguste linearibus intus villosis, cor. st. foliolorum lobis lateralibus ligulatis primario haud duplo brevioribus.——Ad locum "Khurie pass" dictum; *Royle.*

The corolla of this species is from two to three inches long, and except the ventricose base is nearly cylindrical: this readily distinguishes it from *C. elegans*, which is nearly allied in other respects.—(*R. W.*)

7. C. *longifolia* (Wall.:) pubescens volubilis, radice fibrosa, foliis anguste lanceolatis (3–7 uncias longis unamque latis) valde attenuatis, pedunculis plurifloris folio brevioribus, corolla basi ventricosa, tubo clavato, limbi segmentis tubo subdimidio brevioribus sursum latioribus ciliatis, cor. st. foliolorum lobis hirsutis lateralibus angustis ligulatis primario subdimidio brevioribus.—*Wall.! pl. Asiat. rar.* 1. *p.* 56. *t.* 73.——In Nepal.—(*R. W.*)

8. C. *angustifolia* (Wight:) pubescens volubilis, foliis anguste lanceolatis (2–4 uncias longis vix semiunciam latis) valde attenuatis, pedunculis plurifloris, corolla vix basi ventricosa, tubo subcylindraceo, limbi segmentis sursum latioribus tubo multo brevioribus, cor. st. foliolorum lobis anguste ligulatis lateralibus hirsutis primario subdimidio brevioribus.—Ceropegia, *Wall.! asclep. n.* 8, 13, 19.——In Silhet.

Nearly allied to *C. longifolia*, but, so far as I can judge from rather imperfect specimens, it is certainly distinct.—(*R. W.*)

** *Cor. st. foliorum lobi laterales ligulati, gynostegio breviores.*

9. C. *sphenanantha* (W. & A.:) volubilis glabra, foliis distantibus lanceolatis attenuatis, pedunculis paucifloris, calyce minuto laciniis filiformibus, corolla basi ventricosa, tubo clavato, limbo hemisphærico segmentis sursum latioribus longe ciliatis, cor. st. foliolorum lobis ligulatis lateralibus gynostegio brevioribus hirsutis.—*Wall.! Asclep. n.* 6; *Wight.! cat. n.* 1513.——Ad Neelgherry.

10. C. *lanceolata* (Wight:) volubilis glabra, foliis anguste lanceolatis tenuiter acuminatis, pedunculis multifloris, laciniis calycinis filiformibus basin corollæ ventricosam excedentibus, cor. tubo clavato, limbo dilatato segmentis tubo multo brevioribus, cor. st. foliolorum lobis primariis gynostegium paullo superantibus, lateralibus brevioribus acutis ciliatis, folliculis gracilibus glabris.—Ceropegia, *Wall. Ascl. n.* 12.——In Nepal.—(*R. W.*)

11. C. *pusilla* (W. & A.:) herbacea glabra erecta (2–4 uncias alta), radice tuberosa, foliis linearibus, floribus solitariis ascendentibus, corolla basi ventricosa, tubo cylindrico, limbi segmentis tubo multo brevioribus, cor. st. fo-

liolorum lobis gynostegio primariis paullo longioribus, lateralibus paullo brevioribus.—*Wight. cat. n.* 1514.—Ceropegia, *Wall. Asclep. n.* 7.——Ad Neelgherry.

The wild hogs on the hills are said to be very fond of the tubers of this plant.

*** *Cor. st. foliol. lobi laterales acuminati primario approximati, folioli eorumque laborum lateralium apices sinu lato interjecto.*

12. C. *bulbosa* (Roxb.:) volubilis glabra carnosula, radice tuberosa, foliis ex suborbicularibus in lanceolata acuminata, pedunculis folio brevioribus plurifloris, laciniis calycinis corollæ basi ventricosa multo brevioribus, cor. tubo subclavato, limbi segmentis sursum latioribus tubo multo brevioribus ciliatis, cor. st. foliolorum lobis primariis subulatis apice incurvis lateralibus minutis acuminatis primario incumbentibus.—*Roxb. Cor.* 1. *p.* 11. *t.* 7; *fl. Ind.* 2. *p.* 28; *Willd. sp.* 1. *p.* 1275; *Roem. et Sch. syst. veg.* 6. *p.* 2; *Spr. syst.* 1. *p.* 842; *Wall. Ascl. n.* 15, 16; *Wight. in Hook. bot. misc.* 2. *p.* 99, *suppl. t.* 2; *cat. n.* 1515.——Ad Samulcottah, Negapatam, Allahabad, &c.

13. C. *acuminata* (Roxb.:) herbacea volubilis glabra carnosula, radice tuberosa, foliis lineari-lanceolatis apice attenuatis (2–4 uncias longis ac 4–6 lineas latis), pedunculis paucifloris, corolla basi ventricosa, tubo subclavato, limbi segmentis tubo multo brevioribus, cor. st. foliolorum lobis primariis subulatis, lateralibus minutis.—*Roxb. Cor.* 1. *p.* 12. *t.* 8; *fl. Ind.* 2. *p.* 29; *Willd. sp.* 1. *p.* 1276; *Roem. et Sch. syst. veg.* 6. *p.* 2; *Spr. syst.* 1. *p.* 842.——Ad Samulcottah, *Roxburgh.*

There are no specimens of this among Dr Wallich's plants, but, so far as we can judge from the figure and description, it only differs from *C. bulbosa* in the shape of the leaves, and ought rather perhaps to be considered a variety than a distinct species.

**** *Cor. st. foliol. lobi laterales brevissimi obtusi vel primario omnino cohærentes.*

14. C. *Arnottiana* (Wight:) volubilis glabra, caulibus gracilibus, foliis anguste lanceolatis mucronulatis, pedunculis brevibus multifloris, laciniis calycinis anguste subulatis basi corollæ ventricosa longioribus, cor. tubo brevi sursum paullo ampliato, limbo segmentis filiformibus conniventibus tubo longioribus, cor. st. foliolorum lobis primariis ligulatis, lateralibus brevibus obtusis approximatis primarii dorso fere usque ad apices sinu parvo discretos adnatis subciliatis.—Ceropegia, *Wall. Asclep. n.* 3.——In collibus Prome.—(*R. W.*)

15. C. *Wallichii* (Wight:) erecta, radice tuberosa, foliis ovatis sessilibus versus caulis apicem approximatis ciliatis, pedunculis lateralibus brevibus 1-floris, laciniis calycinis subulatis, corolla basi ventricosa, tubo sursum ampliato, limbi segmentis sursum latioribus conniventibus dense ciliatis, cor. st. foliolorum lobis primariis filiformibus, sublateralibus brevibus obtusis ciliatis, folliculis longissimis glabris.—*Wight in bot. mag. tab.* 3267.—C. erecta, *Wall.!?* *Asclep. n.* 1 (*spec. valde manca*).—Nepal; *Wallich* (*in herb. Banks, spec. bon.*) Himalaiah; *Royle.* Prope Lahooghat in Kamaon; *Lindsay* (*in herb. Arnott.*) —(*R. W.*)

16. C. *pubescens* (Wall.:) herbacea volubilis, foliis ovatis tenuiter acuminatis utrinque subpubescentibus, pedunculis folio brevioribus multifloris, corolla cylindracea basi haud ventricosa, limbi segmentis basi latioribus subcordatis sursum attenuatis conniventibus, cor. st. foliolorum lobis primariis subulatis apice recurvis, lateralibus primarium marginantibus sive omnino adnatis.—*Wall.! pl. As. rar.* 2. *p.* 81. *t.* 187; *Asclep. n.* 10.—Ceropegia, *Wall.! Ascl. n.* 17, *a.*—Nepal, *Wallich*. . . . . . ., *herb. Heyne.*—(*R. W.*)

17. C. *tuberosa* (Roxb.:) herbacea glabra volubilis, foliis ex suborbiculatis in ovalia vel ovata cuspidata ac etiam lanceolata tenuiter acuminata, pedun-

culis sæpius geminis pauci vel multifloris folio longioribus brevioribusve, calyce parvo laciniis subulatis, corolla basi ventricosa, tubo sursum ampliato, limbi segmentis anguste sublinearibus villosis tubo dimidio brevioribus marginibus approximatis, gynostegio stipitato, cor. st. foliolorum lobis primariis ligulatis, lateralibus primarium marginantibus eive omnino adnatis, folliculis gracilibus.—*Roxb. Cor.* 1. *p.* 12. *t.* 9; (*in fl. Ind. omissa*); *Willd. sp. pl.* 1. *p.* 1275; *Roem. et Sch.* 6. *p.* 3; *Spr. syst.* 1. *p.* 842; *Wight.! cat. n.* 1516.—Ceropegia, *Wall.! Asclep. n.* 17, *b. c, et* 18.—C. mucronata, *Roth. nov. sp. p.* 179; *Roem. et Sch. l. c. p.* 10.—C. biflora, *Linn.?*—C. Candelabrum, *herb. Rottl.!; Roxb. fl. Ind.* 2. *p.* 27; *Willd. sp.* 1. *p.* 1275.——In ore Coromandeliana haud infrequens. In arenosis prope Negapatam in Tanjore.

This is a very variable species as far as regards the foliage, but is constant in the shape of the flowers, the slender segments of the limb with their margins approximated their whole length and forming a straight beak to the corolla, and in the stalked gynostegium or column of fructification. As Willdenow's and Roxburgh's specimens of *C. Candelabrum* were sent them by Rottler and Klein, we are enabled, on the authority of one, also from Rottler's herbarium, to refer that synonym here; perhaps Roxburgh himself became satisfied of this, as he inserts *C. Candelabrum* in the flora Indica, but omits his own *C. tuberosa.* We likewise refer here to Linnæus' *C. biflora*, but we do so with doubt, because it rests on the authority of an imperfect drawing of Hermann's, which, although we have seen, we cannot positively determine whether or not it belongs to the present species, although we incline to that opinion. We have therefore adopted Roxburgh's name.

Species vix nota.

18. C. *Candelabrum* (Linn.)—*Linn. sp. p.* 309.—*Rheed. Mal.* 9. *t.* 16.

Taken up by Linnæus from Rheede's figure: it approaches very closely to *C. tuberosa*, but presents a considerable difference of habit. From Moon, in his catalogue of Ceylon plants, applying to it the English name of "Common Ceropegia," it would appear to be frequent in that island; but we suspect that, in every instance, modern botanists have had *C. tuberosa* in view. There is a coloured drawing but no specimen in Smith's herbarium, but it may possibly have been copied, although little altered, from Rheede's figure, and coloured from his description.

II. CARALLUMA. *R. Brown.*

Corolla rotata, profunde 5-fida. Gynostegium exsertum. Corona staminea simplici serie 10-loba; lobis 5 staminibus oppositis carnosis antheris incumbentibus; alternis ligulatis apice bipartitis. Antheræ apice simplices. Massæ pollinis erectæ tetragonæ, hinc margine pellucido. Stigma muticum. Folliculi graciles, læves. Semina comosa.—Herbæ erectæ carnosæ aphyllæ. Caules tetragoni, secus angulos dentati. Flores versus ramorum apices. Pedunculi solitarii, uniflori, e ramorum dentium axillis orti.

The crown in this genus, as in *Ceropegia*, consists of 5 3-lobed leaves; the lateral lobes are united by pairs into 5 ligulate forked leaflets, alternate with the primary lobes and with the stamens. So far as we have had opportunities of examining, the same structure, but often curiously modified, is found in *Stapelia* and in all the other genera separated from it. In all, likewise, the pollen masses have a pellucid margin, although not equally conspicuous; on which account, it seems necessary, either to subdivide this group still further, according to the modifications of the corona, or to reunite most of the genera that have been already separated. We have adopted the former of these with regard to the Indian species.

1. C. *adscendens* (R. Brown:) floribus sæpius nutantibus, corollæ segmentis acuminatis glabris.—*Br. in Wern. soc. mem.* 1. *p.* 25; *Roem. et Sch. syst. veg.* 6. *p.* 49; *Spr. syst.* 1. *p.* 842; *Wall.! Asclep. n.* 22; *Wight.! cat. n.* 1517.—Stapelia adscendens, *Roxb. Cor.* 1. *t.* 30.——In aridis Peninsulæ.

2. C. *fimbriata* (Wall.:) floribus nutantibus vel ascendentibus, corollæ segmentis lineari-oblongis cuspidatis fimbriatis.—*Wall. pl. As. rar.* 1. *p.* 7. *t.* 8; *Wight.! cat. n.* 1518.——In Ava, inter ruinas urbis Pagamew, et in collibus ad Yenangeum; *Wallich.* In aridis Peninsulæ; *Wight.*

The flowers in our specimens are much smaller than those in the figure and description given by Dr Wallich; but as we can perceive no other mark of distinction, we have no hesitation in considering them the same species.

III. BOUCEROSIA. *W. & A.*

Corolla subcampanulata, 5-fida; segmentis late triangularibus, sinu acuto. Gynostegium vix exsertum. Corona staminea gamophylla, 15-loba, duplici serie; lobis 5, interioribus staminibus oppositis antheris incumbentibus; cæteris exterioribus erectis vel apice subincurvis interioriorum dorso adhærentibus. Antheræ apice simplices. Massæ pollinis erectæ tetragonæ, hinc margine pellucido. Stigma muticum. Folliculi læves teretes apice attenuati. Semina comosa.—Plantæ carnosæ, aphyllæ, erectæ, tetragonæ, angulis dentatis. Flores numerosi, terminales, umbellati.

In *Caralluma* the lateral lobes of each leaflet of the corona are separted by means of the middle or primary lobe, with the edge of which they cohere at the base: in *Boucerosia*, on the contrary, the margins of the lateral lobes of each leaflet cohere, and their line of junction is also attached along the middle of the back of the primary lobe. In both genera the lateral lobes cohere by the margins with the lateral ones of the adjoining leaflet; so that in *Caralluma* we have a corona of 10 narrow and 5 broad segments attached to each other by the margins only, and forming, consequently, a simple series; while, in *Boucerosia*, the 5 broad segments are surrounded by the 10 narrow ones, and a double corona is formed.

1. B. *umbellata* (W. & A.:) corollæ segmentis glabris.—*Wight.! cat. n.* 1519. Caralluma umbellata, *Haworth. syn. pl. succul. p.* 47; *Spr. syst.* 1. *p.* 842; *Wall.! Asclep. n.* 20, 21.—Stapelia umbellata, *Roxb. cor.* 3. *t.* 241; (*in fl. Ind. omissa*).—S Cullamulia, *Ham.*——In provinciis Peninsulæ australibus in aridis frequens.

2. B. *crenulata* (W. & A.:) corollæ segmentis longe fimbriatis.—Caralluma crenulata, *Wall. pl. As. rar.* 1. *p.* 6. *t.* 7.—In arenosis montium prope urbem Segain, ad ripam dextram fluminis Irawaddi, contra Avam situm; *Wallich.*

Dr Wallich describes the corona as simple, and the leaflets with denticuli at their base: by comparing his figure, however, with *B. umbellata*, these denticuli obviously are the lateral lobes, and form the exterior series.

IV. HUTCHINIA. *W. & A.*

Corolla subcampanulata 5-fida. Gynostegium vix exsertum. Corona staminea gamophylla, 20-loba; lobis 5 interioribus antheris incumbentibus; cæteris ternis subexterioribus interioribus (seu primariis) alternantibus, duobus lateralibus intermedium superantibus. Antheræ apice simplices. Massæ pollinis erectæ hinc margine pellucido.—Plantæ carnosæ, aphyllæ, erectæ, ramosæ, tetragonæ, angulis dentatis. Flores numerosi, terminales, subumbellati. Corolla intus pilosa.

We dedicate this genus to Mr Hutchin of Norwich, a most successful cultivator and liberal distributor of succulent plants of all kinds, notwithstanding he has laboured under the combined disadvantages of confined space, and residence in the centre of a large town. The similarity of spelling between this name and that of *Hutchinsia* can be no obstacle, the addition of the consonant in the latter giving so very different a sound, that no confusion can possibly arise.

1. H. *Indica* (W. & A.)—*Wight.! cat. n.* 1520.——In Peninsulæ aridis (forsan montium australium).

A very branched plant, 4–6 inches high: branches somewhat diffuse, 4-sided, with numerous reflexed teeth on the angles. Flowers pale, greenish-white, outwardly tinged with purple, and covered with purplish hairs within. We have not seen the follicles.

V. ERIOPETALUM. *Wight.*

Corolla subcampanulata, 5-fida; segmentis anguste linearibus, sinu lato. Corona staminea gamophylla, 15-loba; lobis 5 interioribus antheris incumbentibus; cæteris exterioribus erectis approximatis seriei interiori adhærentibus. Antheræ apice simplices. Massæ pollinis juxta basin affixæ, erectæ, stigmati incumbentes, hinc margine pellucido. Folliculi graciles, glabri.—Herbæ erectæ, subramosæ. Caules teretes. Folia minuta, squamiformia, adpressa. Flores parvi, umbellati; umbellæ laterales aut terminales, sessiles. Corollæ segmenta tubo longiora, sæpius intus dense villosa aut lanata; hinc nomen.

This genus agrees in habit with *Microstemma* of Brown, but differs in the form of the crown: on the other hand, it resembles *Boucerosia* in the corona, but has a widely different habit.

1. E. *parviflorum* (Wight:) corollæ segmentis tubo vix longioribus lanatis, cor. stam. gynostegium superante.——Doongie; *Royle.*—(*R. W.*)

2. E. *attenuatum* (Wight:) corollæ segmentis gracilibus attenuatis villosis tubo duplo triplove longioribus, cor. staminea gynostegio breviore.——Doongie; *Royle.*—(*R. W.*)

3. E. *lævigatum* (Wight:) corollæ segmentis linearibus margine undulatis fere glabris tubo subduplo longioribus, corona staminea gynostegium æquante.—Gomphocarpus lævigata, *Ham.*—Microstemma, *Wall.! Asclep. n.* 23.——In sylvis Cossalæ, Gorukpur; *Hamilton.*—(*R. W.*)

VI. HOYA. *R. Brown.*—Sperlingia, *Vahl.*

Corolla rotata, 5-fida. Corona staminea 5-phylla, foliolis depressis patentibus carnosis, angulo interiore producto in dentem antheræ incumbentem. Antheræ membrana terminatæ. Massæ pollinis basi affixæ, conniventes, compressæ. Stigma muticum vel subapiculatum. Folliculæ læves. Semina comosa.—Frutices vel suffrutices volubiles scandentes aut decumbentes. Folia opposita, carnosa, vel membranacea. Umbellæ laterales, multifloræ.

§ I. *Volubiles scandentes aut decumbentes, sæpe radicantes. Folia carnosa. Coronæ st. foliola subtus sulcata.*—Hoyæ veræ.

* *Cor. st. foliola planiuscula, late obovata aut suborbicularia.*

1. H. *velutina* (Wight:) volubilis, foliis ovalibus breviter acuminatis margine revolutis supra glabris subtus velutinis, corolla extus pubescente intus glabra segmentis latis margine revolutis, cor. st. foliolis suborbicularibus depressis supra nitidulis angulo interiore obtuso stigmati umbilicato incumbente. —Hoya, *Wall.! Asclep. n.* 39.

No locality is attached. This species approaches very closely, both in habit and size of the flowers, to the *Corona Ariadnes seu Sussuela* of Rumphius, Amb. 5. t. 172; but Rumphius says, that the leaves of his plant are glabrous; perhaps he may have overlooked the pubescence on the under side. Roxburgh, in his flora Indica, 2. p. 31, refers Rumphius' figure to his *Asclepias Sussuela*, but they cannot be the same plant or perhaps same genus: Roxburgh's is said to have the flowers, when expanded, 3 inches across; I have seen no Indian Asclepiadeæ of such a size, unless, as may be the case, Roxburgh alludes to the umbels and not to a single flower.—(*R. W.*)

2. H. *orbiculata* (Wall.:) volubilis, foliis carnosis orbicularibus glabris, pedunculis folio brevioribus multifloris, corolla intus pubescent segmentis acutis, cor. st. foliolis obovatis suborbicularibusve depressis angulo anteriore obtuso, stigmate apiculato.—*Wall.! Asclep. n.* 32.—*Rumph. Amb.* 5. *t.* 175. *f.* 28. ——Apud Prome, in Ava; *Wallich.*

There is in my specimen usually a verticil of 4 leaves at each joint; 2 of them, smaller, are the proper leaves of the stem; two, larger, are from the axil of one of the others.—(*R. W.*)

3. H. *pendula* (W. & A.:) volubilis, foliis carnosis glabris ex oblongo-ovalibus acutis in late ovata acuminata margine revolutis, pedunculis petiolum paullo superantibus pendulis multifloris, corolla intus pubescente, cor. st. foliolis obovalibus obtusissimis depressis angulo interiore brevi apice truncato, stigmate apiculato.—α, *Rheedei;* foliis oblongo ovalibus acutis.—Asclepias pendula, *Roxb. fl. Ind.* 2. *p.* 36; *in cœt. merc. Ind. or. mus. tab.* 612.—*Rheed. Mal.* 9. *t.* 13.—β, *Neelgherrense;* foliis ovatis acuminatis.—*Wight.! cat. n.* 1521.—H. revoluta, *Wight.! in Wall. herb. soc. Linn.*—Hoya, *Wall.! Asclep. n.* 27, 41?—α. In montibus Circars; *Roxburgh.* Malabar; *Rheede.*—β. Ad Neelgherry; *Wight.*

There are some specimens in Finlayson's herbarium (Wall. asclep. n. 41), but not in flower, very similar to the Neelgherry plants, and have the margin of the leaves folded back in the same way; but the leaves are much smaller, and the stem throws out roots, so that they can scarcely be referred to the same species. Our Neelgherry specimens are also imperfect. With regard to Roxburgh's plant, it must be confessed that his specific character, "nectaries protruding five horns at the base," does not appear to be applicable to ours; but, on the other hand, his after description, "nectary stellate; O may represent one of its five parts much magnified," is in accordance, nor does he there notice the horns formerly mentioned. In his drawing at the India house, one of the dissections agrees well with our species, but there is another which represents the coronal leaflet expanding at the back into a long tapering horn, and which we scarcely understand; we almost suspect that the artist had here taken a side-view of the leaflet, without regard to perspective, thus representing a kind of longitudinal section, and that Roxburgh has trusted to the accuracy of the figure, and described in consequence, as part of his specific character, the projecting horns, which he himself had not seen. There is a specimen in the Banksian herbarium, either named by Roxburgh, or supposed to be the same with his plant, to which Mr Brown has attached the name of *H. Nicobarica* (Br. mst.); but Roxburgh has expressly mentioned that his plant is a "native of the mountainous part of the Circars," while that named by Mr Brown seems to have come from the Nicobar islands, to the north of Sumatra: although we saw this specimen, yet unfortunately we neither examined it with attention, nor made any notes upon it; if, however, our recollection serves us rightly, it appeared more allied to, if not the same with *H. Hookeriana* of Wight.

4. H. *longifolia* (Wall.:) volubilis radicans, foliis carnosis anguste lanceolatis acuminatis basi attenuatis (3–6 uncias longis et 6–12 lineas latis), corolla intus pubescente margine ciliata, cor. st. foliolis late obovatis obtusis angulo interiore acuminato stigmate incumbente.—*Wall.! Asclep. n.* 40.—Hoya, *Wall.! Asclep. n.* 44?—Napal; Singapore?; *Wallich.* Apud Mussooree, Himilaiah; *Royle.*

The Singapore specimen (n. 44.) is not in flower.—(*R. W.*)

5. H. *linearis* (Wall.:) volubilis pilosa, caule tenui folioso, foliis carnosis anguste linearibus acutiusculis, corolla intus subpubescente margine ciliata segmentis obtusis, cor. st. foliolis suborbicularibus angulo interiore stigmati

incumbente.—*Wall.! Asclep. n.* 34; *Don. prod. fl. Nep. p.* 130.——Nepal; *Wallich.*—(*R. W.*)

6. H. *parviflora* (Wight:) volubilis glabra, ramis tenuibus, foliis carnosis lanceolatis basi apice attenuatis, pedunculis gracilibus folio subæquantibus multifloris parvifloris, corolla glabra, cor. st. foliolis late obovatis emarginatis obcordatisve angulo interiore acuminato stigmati apiculato incumbente.—Hoya, *Wall.! Asclep. n.* 33.——Maulmyne; *Wallich.*—(*R. W.*)

** *Cor. st. foliola supra planiuscula ovata vel lanceolata.*

7. H. *fusca* (Wall.:) volubilis glabra, foliis coriaceis vix carnosis lineari-oblongis acuminatis (5–8 uncias longis 1–1½ latis) transversim venosis, pedunculis brevibus multifloris, corolla glabra segmentis obtusis, cor. st. foliolis ovatis obtusis angulo interiore attenuato erecto stigma apiculatum superante.—*Wall.! pl. As. rar.* 2. *p.* 78. *t.* 175; *Asclep. n.* 30.——Nepal; Silhet; *Wallich.* The flowers are rather small, brownish coloured, and very numerous.—(*R. W.*)

8. H. *Hookeriana* (Wight:) volubilis glabra, foliis carnosis oblongo-lanceolatis 3-nerviis (6–10 uncias longis 1–1½ latis), pedunculis brevibus multifloris, corolla (parvula) glabra segmentis obtusis, cor. st. foliolis ovatis acutis angulo interiore brevi, stigmate obtuso.—Hoya, *Wall.! Asclep. n.* 28, 39 (ex parte).——Chittagong; Silhet; *Wallich.*—(*R. W.*)

9. H. *parasitica* (Wall.:) parasitica scandens radicans glabra, ramis tenuibus teretibus, foliis carnosis glabris nitidis tenuiter basi trinerviis oblongo-lanceolatis attenuatis, pedunculis folio subduplo brevioribus multifloris, corolla glabra alte 5-fida, cor. st. foliolis ovatis acutis angulo interiore stigmati incumbente.—*Wall.! Asclep. n.* 29.—Asclepias parasitica, *Roxb. fl. Ind.* 2. *p.* 42; *in cœt. merc. Ind. or. mus. tab.* 1382.——In sylva Sunderbund ad æstuarium Gangeticum; *Roxburgh.*—(*R. W.*)

10. H. *ovalifolia* (W. & A.:) volubilis radicans glabra, foliis carnosis ovalibus utrinque acuminatis, pedunculis folio brevioribus multifloris, corolla intus puberula segmentis ovatis acutis, cor. st. foliolis ovalibus obtusis angulo interiore brevi, stigmate mutico.—Hoya, *Wight.! cat. n.* 1522; *Wall.! Asclep. n.* 31, 43?, 45?——Ad Neelgherry. Singapore?; *Wallich.*
The Singapore specimens are not in flower, and are probably different. This is very nearly allied to *H. parasitica*, but apparently quite distinct.

11. H. *Arnottiana* (Wight:) volubilis glabra radicans, foliis carnosis late ovatis acuminatis, corolla intus puberula segmentis orbicularibus cuspidatis, cor. st. foliolis ovalibus subacuminatis angulo interiore brevi, stigmate mutico.—Hoya, *Wall.! Asclep. n.* 35.——In Nepal; *Wallich.*—(*R. W.*)

12. H. *gymnanthera* (Wight:) aphylla?, caulis epidermide lævi nitido, umbellis multifloris breve pedunculatis parvifloris, cor. st. foliolis ovatis, antherarum membranaceis apicibus pellucidis angustissimis pollinis massas vix tegentibus.—*Wall.! Asclep. n.* 142 (absque nomine).—........; *herb. Finlayson.*
The specimens of this plant which I have seen are leafless, but whether from accident or otherwise, I have not at present the means of ascertaining.—(*R. W.*)

*** *Cor. st. foliola supra convexa ovalia vel ovata margine revoluta.*

13. H. *Brunoniana* (Wight:) volubilis glabra, ramis obsolete tetragonis, foliis carnosis ovali-oblongis acuminatis, corolla subcarnosa extus glabra intus barbata, cor. st. foliolis ovalibus acutis supra convexis margine revolutis.—Hoya, *Wall.! Asclep. n.* 37.—.............*herb. Wallich.*
The hairs on the inside of the corolla, in the dried specimens, have the appearance of collapsed hollow tubes.—(*R. W.*)

14. H. *carnosa* (R. Brown:) volubilis glabra, caulibus ramisque teretibus, foliis carnosis ovali-oblongis acuminatis, umbellis pendulis breve pendunculatis, pedicellis piloso-pubescentibus, corolla carnosa extus glabra intus omnino barbata, cor. st. foliolis ovatis acutis apice porrecto supra convexis margine revolutis, antherarum membranis majusculis stigma obtuse apiculatum occultentibus, massis poll. hinc pellucide marginatis.—*R. Br. in Wern. soc. trans.* 1. *p.* 27 (cum syn. omn.); *prod. fl. nov. Holl.* 1. *p.* 460; *Roem. et Sch. syst. veg.* 6. *p.* 51; *Spr. syst.* 1. *p.* 843.—Schollia crassifolia, *Jacq.*—Apud Canton in China; *Loureiro.*——" In Asiæ tropicæ variis regionibus;" *R. Brown.*

I have taken upon me to add the character of this species, although I am not aware of it having been found, either in any part of the East India Company's possessions or in Ceylon; nor does it seem to be among the plants in the Botanic Garden of Calcutta. I believe my friend Dr Wight was not acquainted with it when he left with me the description of *H. Brunoniana*, otherwise he might have been induced to unite the two. They indeed differ by characters not, I fear, of a permanent nature: in the mutilated specimen of *H. Brunoniana*, which I have examined, the pedicels seem to be glabrous; the corolla is of a less fleshy nature than in *H. carnosa*, and its segments are not bearded close up to the margin, which is glabrous: the shape of the coronal leaflets also differs slightly; their extremity is (at least by drying) recurved in *H. Brunoniana*, and straight in the other; but they are of the same firm, almost horny, texture: the membranous tip of the anthers is alike in both: and *H. Brunoniana* has the one margin of each pollen mass marked with an elevated pellucid line, exactly as *H. carnosa*. In the former, the leaves are nearly twice the size of those of the latter.—(*G. A. W.-A.*)

15. *H. lanceolata* (Wall.:) volubilis glabra, caulibus tenuibus foliosis, foliis carnosis lanceolatis basi apice acuminatis, pedunculis folio brevioribus pauciforis, corolla intus pubescente segmentis obtusis, cor. st. foliolis supra convexis margine recurvis angulo interiore elongato obtuso erecto stigma apiculatum superante.—*Wall.! Asclep. n.* 36; *Don. prod. fl. Nep. p.* 130.——In Nepal et Silhet; *Wallich.*—(*R. W.*)

**** *Cor. st. foliola elongata triangularia.*

16. H. *macrophylla* (Wight:) volubilis, foliis late ovatis acuminatis basi obtusis subcordatisve 3-nerviis carnosis glabris (7–10 uncias longis 4–10 latis), pedunculis multifloris, corollæ segmentis obtusis cuspidatis, cor. st. foliolis corollam fere æquantibus lineari-triangularibus coriaceis angulo interiore tereti erecto obtuso, folliculis subulatis tenuibus.—*Wall.! Asclep. n.* 138 (absque nom.).——Penang; *Wallich; Finlayson.*

The specimens of this most magnificent species have unfortunately very few flowers on them, and these are so much injured (the pistillum being entirely destroyed), that it is impossible to say with certainty if it belong truly to the genus.—(*R. W.*)

Species floribus adhuc incognitis altamen Hoyis veris *adscribendæ.*

17. H. *Finlaysonii* (Wight:) volubilis glabra, ramis teretibus, foliis oblongo-lanceolatis obtuse acuminatis supra (siccitate) reticulatis subtus subaveniis (5–7 uncias longis 1½–2 latis) breve petiolatis.—Hoya, *Wall.! Asclep. n.* 42. ——Penang? ; *Wallich* et *Finlayson.*

No station is attached to the specimens, but from the same plant existing in Wallich's and Finlayson's herbaria, it is probably from Penang.—(*R. W.*)

18. H. *obtusifolia* (Wight:) volubilis glabra, foliis carnosis ovali-oblongis basi apice obtusis cuspidatis supra nitidis subtus transversim (siccitate) obscure venosis, pedunculis folio subdemidio brevioribus multifloris.—Hoya, *Wall.! Asclep. n.* 38.——..............; *Wallich.*

It is probable, from the transverse venation, which, however, can only be seen in the dry state, that this species ranks near *H. fusca*. Judging from the appearance of the peduncles, the flowers are very numerous, but fall off very readily.—(*R. W.*)

§ 2. *Folia membranacea. Cor. st. foliola subtus esulcata.*

19. H. *viridiflora* (R. Brown :) volubilis, foliis ovatis cordatisve acuminatis membranaceis utrinque glabris, corolla glabra segmentis ovatis acutiusculis, cor. st. foliolis supra planis obovalibus obtusissimis angulo interiore brevi obtuso, folliculis divaricatis crassis obtusis ferrugineo-farinosis.—*Br. in Wern. soc. mem.* 1. *p.* 26; *Roem. et Sch. syst. veg.* 6. *p.* 51; *Spr. syst.* 1. *p.* 843; *Wall.! Asclep. n.* 24 (excl. *f*), 26, 60, *c* (ex parte); *Wight! in Hook. bot. misc.* 2. *p.* 98. *suppl. t.* 1; *cat. n.* 1523.—Asclepias volubilis, *Linn. suppl.* 170; *Willd. sp. pl.* 1. *p.* 1279; *Roxb. fl. Ind.* 2. *p.* 36.—A. viridiflora, *Roxb. in cat. merc. Ind. or. mus. tab.* 611.—Apocynum tiliæfolium, *Lam. enc. meth.* 1. *p.* 214.—*Rheed. Mal.* 9. *t.* 15.—β, *alpina*; floribus majoribus, corollæ segmentis obtusis.—*Wight.! cat. n.* 1523. *b.*—Hoya, *Wall.! Asclep. n.* 25.——α. In ore Coromandeliana frequens. Apud Morang et Gongachora, *Hamilton*; Silhet, *Wallich.*—β. Ad Neelgherry; *Noton*; *Wight.*

The alpine variety has much larger flowers, and the segments of the corolla very obtuse: it does not appear, however, to differ in any other respect.

20. H. *Lacuna* (herb. Ham.:) volubilis, foliis coriaceo-membranaceis cordatis ovatisve acuminatis supra pubescentibus subtus velutinis, corollæ segmentis obtusis, cor. st. foliolis angulo interiore brevi subulato erecto, folliculis e basi crasso obtuse acuminatis velutinis.—*Herb. Ham.! n.* 770.—*Wall.! Asclep. n.* 24. *f*, 146, 156 (absque nom.).——Korma; *Hamilton.* Segaen; *Wallich.*—Nom. officin. Hind., *Lakun.*—(*R. W.*)

VII. PTEROSTELMA. *Wight.*

Corolla rotata, 5-fida. Corona staminea 5-phylla; foliolis membranaceis lateribus reflexis, angulo interiore in dentem subulatum erectum producto. Antheræ membrana terminatæ. Massæ pollinis erectæ, approximatæ, ad dorsum corpusculi basi affixæ. Stigma apiculatum.—Frutex volubilis? aut decumbens? Folia opposita, oblonga, acuminata, carnosa. Flores majusculi. Corolla alte 5-fida, segmentis lineari-lanceolatis, tubo piloso. Coronæ foliola lata, lateribus arcte reflexis marginibusque conniventibus, ita ut Papilionis sedentis alæ.

This genus is allied to *Hoya* both in habit and in the structure of the flowers, differing principally in having foliaceous in place of fleshy crown-leaves, and in the dorsal not lateral attachment of the pollen masses to the corpuscle.

1. P. *acuminata* (Wight).—*Wall.! Asclep. n.* 129 (absque nomine).——In Silhet; *Wallich.*

My specimens do not possess a perfect umbel.—(*R. W.*)

VIII. PHYSOSTELMA. *Wight.*

Corolla rotata, patens, repande 5-dentata. Corona staminea 5-phylla; foliolis inflatis, antheris oppositis. Antheræ membrana terminatæ. Filamenta e basi ad apicem bipartita, cruribus cum crure filamenti utrinque proximimi arcte coalitis filamenta 5 antheris coronæque foliolis alternantia corpusculis stigmatis opposita efformantibus. Massæ pollinis basi affixæ, erectæ, subcylindraceæ. Stigma depressum.—Frutex volubilis. Folia opposita ovato-lanceolata, acuminata, glabra, carnosa. Umbellæ laterales, longiuscule pedunculatæ, pauciflora. Flores magni.

This curious genus is allied in habit to *Hoya*, but is at once distinguished from it and every other by the remarkable division of its filaments; and, when better known, may tend to change our ideas of the structure of these organs, in this division at least of the order, by rendering it probable that the normal form is to have ten filaments and ten anthers, which become reduced to five by pressure during their development. The principal objection to this view in the present instance arises from there being only five anthers to the ten filaments, but which on further examination may be found to strengthen rather than weaken the conjecture.

1. P. *Wallichii* (Wight).—*Wall.! Asclep. n.* 130 (absque nom.)——Singapore; *Wallich.* . . . . . . ; *Finlayson.*—(*R. W.*)

IX. MARSDENIA. *R. Brown.*

Corolla urceolata, 5-fida, nunc subrotata. Corona staminea 5-phylla; foliolis compressis, intus simplicibus. Antheræ membrana terminatæ. Massæ pollinis erectæ, basi affixæ. Stigma muticum vel rostratum. Folliculi læves. Semina comosa.—Frutices sæpius volubiles. Folia opposita, membranacea, latiuscula. Cymæ vel thyrsi interpetiolares.

The flowers are usually rather small, sometimes glabrous, sometimes villous all over, and sometimes the throat is closed with long white shining hairs; according to which differences we subdivide the genus. The tube is about the length of the column or gynostegium.—*Asclepias plicata*, Roth, nov. sp. p. 176, may probably be a species of *Marsdenia*.

§ 1. *Corollæ faux nuda.*

1. M. *Brunoniana* (W. & A.:) volubilis glabra, foliis late cordato-ovatis acuminatis, pedunculis petiolo brevioribus, floribus cymosis majusculis glabris, corollæ segmentis obtusis, cor. st. foliolis attenuatis gynostegium æquantibus, stigmate obtuse apiculato.—*Wight.! cat. n.* 1524.——Prope Columala.

The cymes in this species are small, but the flowers are the largest of the genus we have met with.

§ 2. *Corollæ tubus intus pubescens.*

2. M. *Roylei* (Wight:) volubilis tota tomentosa, foliis late cordatis ovatisve acuminatis, cymis folio brevioribus, calyce hirsuto, corollæ intus hirsutæ fauce plicis 5 sinubus oppositis coronata, cor. st. foliolis gynostegio multo longioribus apice filiformibus, stigmate obtuse apiculato.—M. velutina, *herb. Royle* (non *Brown*).—Kenour; *Royle.* Simla; *Dom. Comitissa de Dalhousie.*

The plicæ at the bottom of the sinus between the segments of the corolla almost resemble thickish scales.—(*R. W.*)

3. M. *angustifolia* (Wight:) volubilis ramosissima, caule ramis pedunculisque puberulis, foliis adultis glabris subsessilibus lanceolatis, cymis folio brevioribus geminis paucifloris, corollæ segmentis ligulatis tubo intus pubescente, cor. st. foliolis gynostegio brevioribus, stigmate truncato.—Marsdenia, *Wall.! Asclep. n.* 58; *Rumph. Amb.* 5. *t.* 40. *f.* 1?——In collibus apud Prome; *Wallich.*—(*R. W.*)

§ 3. *Corollæ faux atque tubus barbati, tubi pilis 5-fasciculatis reflexis.*

4. M. *tinctoria* (R. Brown:) volubilis, foliis ovatis oblongisve acuminatis basi cordatis subglabris supra versus basin glandula instructis, thyrsis lateralibus, floribus parvis, fauce barbata, cor. st. foliolis gynostegium subæquantibus subulatis, stigmate mutico.—*Br. in Wern. soc. mem.* 1. *p.* 39; *Roem. et Sch.* 6. *p.* 53; *Wall.! Asclep. n.* 56; *Wight! cat. n.* 1525.—M. monostachya, *Wall. mst.*—Pergularia tinctoria, *Spr.* 1. *p.* 844.—Asclepias tinctoria, *Roxb. fl. Ind.* 2. *p.* 43; *in cœt. merc. Ind. or. mus. tab.* 615.—Cynanchum tingens, *herb. Ham! n.* 752.——Silhet et Prome; *Wallich.* Gualpara; *Hamilton.*

Although Hamilton in his herbarium names this *Cyn. tingens*, it is quite distinct from a plant he himself had previously sent from Pegu to the Calcutta Botanic Garden under the name of *Asclep. tingens:* this last is our *Gymnema tingens*. We consider this a very doubtful native of the Peninsula; our only specimens from thence being from Klein's herbarium, without the locality mentioned; while, on the other hand, in Wallich's collection, specimens from the same herbarium are marked as procured from the Missionary garden.

5. M. *Hamiltonii* (Wight:) volubilis glabra, foliis cordato-oblongis acuminatis, cymis subcapituliformibus multifloris petiolo brevioribus, floribus parvis, corollæ tubo ventricoso fauce villis clausa, cor. st. foliolis gynostegio longioribus apice attenuatis, pollinis massis oblongis corpusculo minutissimo affixis, stigmate apiculato.—Cynanchum, *Wall.! Asclep. n.* 84.—C. reticulatum, *herb. Ham.! n.* 750.—Apud Sukanagur; *Hamilton.*

From the mst. synonyms attached to this plant in his herbarium (now belonging to the University of Edinburgh), Hamilton appears to have taken his specific name from the supposition of its being *Apocynum reticulatum* of Willdenow: as, however, it is certainly not so, and as the leaves of this species are much less distinctly reticulated than those of *M. tinctoria,* I have taken the liberty of changing the name.—(*R. W.*)

§ 4. *Corollæ faux nuda, tubus sparsim pilosus. Cor. st. foliola latiuscula, apice truncata vel bifida lobo uno acuminato alteram cariniformam prope basin gerente. Cymæ patentes magnæ multifloræ, floribus majusculis.*

The two plants referred to this section may eventually form a distinct genus, but we can find no good character by which to separate the one from *Marsdenia*, and the other is so closely allied to it, that it can only be distinguished by an examination of the corona.

6. M. *Calesiana* (Wight:) volubilis pubescens, foliis cordatis vel cordato-ovatis acuminatis utrinque tomentosis, cymis (magnis) multifloris, corollæ segmentis angustis acutis margine revolutis, cor. st. foliolis alte bifidis lobo uno longiore antheræ incumbente altero breviore versus basin cariniformi, stigmate apiculato.—Pergularia Calesiana, *herb. Ham.!; Wall.! Ascl. n.* 57, *e.* ——Apud Morang; *Hamilton.*—(*R. W.*)

7. M. *tenacissima* (W. & A.:) volubilis, foliis cordatis acuminatis utrinque tomentosis, cymis magnis, corollæ segmentis latis obtusis, cor. st. foliolis latis apice truncatis subintegris vel bifurcatis, stigmate obtuse apiculato.—*Wight.! cat. n.* 1520.—Asclepias tenacissima, *Roxb.! Cor.* 3. *p.* 35. *t.* 240; *fl. Ind.* 2. *p.* 31; *in cœt. merc. Ind. or. mus. tab.* 1381; *Wall.! Ascl. n.* 57.—A. tomentosa, *herb. Madr.!*—A. echinata, *herb. Madr.!*—Gymnema tenacissima, *Spr. syst.* 1. *p.* 844.——In collibus Rajemahl; *Roxburgh.*—Chittagong; *Wallich.*

Although we have it from the Madras herbarium, it is probably not a native of the Peninsula: the specimens under the name of *A. tomentosa* are from the Missionaries' garden; that under *A. echinata* was sent to Klein so named by Heyne, but is obviously not the plant of Roxburgh.

X. COSMOSTIGMA. *Wight.*

Corolla rotata, 5-partita. Corona staminea 5-phylla; foliolis compressis bifidis. Antheræ ventricosæ membrana terminatæ. Massæ pollinis erectæ, corpusculo stigmatis ope appendiculi longi penduli basi affixæ. Stigma muticum ala angusta flexuosa repanda coronatum. Folliculi magni lineari-oblongi obtusi læves. Semina comosa.—Frutices volubiles. Folia opposita, membranacea. Corymbi sæpe geminati, longe pedunculati. Flores parvi.

I have separated this from *Marsdenia* principally on account of the very peculiar and beautiful stigma: the cleft leaflets of the corona allies it to the last section of that genus.

1. C. *racemosa* (Wight).—Asclepias racemosa, *Roxb. fl. Ind.* 2. *p.* 32; *in cat. merc. Ind. or. mus. tab.* 1818.—Nerium piscidium, *herb. Wall.! n.* 59, *ex hort. bot. Calc.* (non *Roxb.*)——Silhet et Chittagong; *Wallich.*—(*R. W.*)

XI. HETEROSTEMMA. *W. & A.*

Corolla rotata, 5-partita. Corona staminea 5-phylla; foliolis dilatatis intus processu auctis. Antheræ membrana stigmati incumbente terminatæ. Massæ pollinis erectæ, obsolete tetragonæ, hinc margine pellucida. Folliculi læves. Semina comosa.—Frutices volubiles. Folia opposita, supra versus basin glandula minuta instructa. Umbellæ parvæ, interpetiolares, breve pedunculatæ.

Owing to the variable nature of the corona in this genus, we have been under the necessity of making its essential character rest on the form and structure of the pollen masses (which resemble those of *Stapelia* and its allies), and the twining habit. Each of the species referred to this genus has a form of the crown peculiar to itself: the first has the leaflets dilated, entire, and spreading, with the interior process similar in form but smaller; the second has them obcordate, much dilated, with the margins free and overlapping, and the interior process erect fleshy, incumbent on and nearly equalling the gynostegium in height; in the third they are furnished with a subulate tooth or horn on the back, while the interior process is reduced to a mere gland.

1. H. *Tanjorensis* (W. & A.:) volubilis glabra, foliis late ovatis oblongisve breviter acuminatis basi obtusis cordatisve (2–4-uncias longis 1–2 latis), pedunculis petiolo brevioribus paucifloris, cor. st. foliolis patentibus latis truncatis processu interiore linguiformi, folliculis divaricatis tenuibus glabris apice uncinatis.—*Wight! cat. n.* 1527.—Stapelia involucrata, *herb. Madr.!; Wall.! Asclep. n.* 112.——In arenosis provinciæ Tanjore.

We have met with this plant in collections, under the name of *Stapelia volubilis* and *S. involucrata,* sent by the Tranquebar missionaries, but both of these names are equally applicable to the other species.

2. H. *Wallichii* (Wight:) volubilis, ramis utrinque linea laterali pubescente, foliis ovatis acuminatis glabris (4–8 uncias longis 2–3 latis), pedunculis brevissimis paucifloris, cor. st. foliolis valde dilatatis obcordatis processu interiore erecto carnoso gynostegio incumbente.—*Wall.! Asclep. n.* 154 (absque nom.) ——Sheopore; *Wallich.*

The interior processes seem to cohere slightly by the margins, forming as it were a tube around the gynostegium, but, from the imperfection of my specimens, I am far from certain that they actually do so.—(*R. W.*)

3. H. *alata* (Wight:) volubilis, ramis utrinque linea laterali pubescente, foliis ovatis acuminatis glabris (2–5 uncias longis 1–3 latis), umbellis paucifloris subsessilibus, cor. st. foliolis cuspidatis dorso dente subulato instructis processu interiore glandulæformi, folliculis tenuibus glabris divaricatis.—Hoya alata, *Wall.! Asclep. n.* 136.——Nepal; *Wallich.*—Massooree; *Royle.*—(*R. W.*)

XII. PERGULARIA. *Linn.; R. Brown.*

Corolla hypocrateriformis, tubo urceolato. Corona staminea 5-phylla; foliolis compressis apice indivisis, intus lacinula auctis. Antheræ membrana terminatæ. Massæ pollinis erectæ. Stigma muticum. Folliculi læves, ventricosi. Semina comosa.—Plantæ volubiles. Folia latiuscula, membranacea. Cymæ interpetiolares.—Flores flavescentes, odorati.

1. P. *pallida* (W. & A.:) volubilis, ramis tenuibus molliter pubescentibus, foliis cordatis acuminatis, cymis subsessilibus multifloris, corollæ segmentis ligulatis tubo intus glabro gynostegium vix superante longioribus, cor. st. ejusque processu stigma apiculatum superantibus.—*Wight.! cat. n.* 1528.—P. viridis, *herb. Ham.!*—P. limbata, *Wall.! Asclep. n.* 61.—Asclepias pallida,

Roxb. fl. Ind. 2. *p.* 48.——Courtallum (vix tamen indigena); *herb. Wight.* Gongachora et Munghir; *Hamilton.* Burma et Penang; *Wallich.* Himalaiah; *Royle.*

2. P. *odoratissima* (Smith:) volubilis, ramis molliter pubescentibus, foliis cordatis acuminatis ad venas pubescentibus, cymis breve pedunculatis multifloris, corollæ segmentis brevibus obtusis, tubo gynostegium duplo superante intus lineis quinque deflexo-pilosis e sinubus decurrentibus, cor. st. gynostegium æquante.—*Sm. ic. pict. t.* 16; *Roxb. in cœt. merc. Ind. or. mus. t.* 122; *Brown in Wern. soc. mem.* 1. *p.* 31; *Andr. bot. rep. t.* 185; *Roem. et sch. syst. veg.* 6. *p.* 54; *Spr. syst.* 1. *p.* 843; *Wall.! Asclep. n.* 60, *a, c* (ex parte); *Wight.! cat. n.* 1529.—Cynanchum odoratissimum, *Lour. fl. Cochin.* 1. *p.* 206. —Asclepias odoratissima, *Roxb. fl. Ind.* 2. *p.* 47.—*Rumph. Amb.* 7. *t.* 26. *f.* 1. ——In Peninsula (vix ac ne vix indigena); *herb. Madras.* Prome; *Wallich.* Pagamew; *Wallich.* Patma; *Hamilton.* China? *Finlayson.*

P. minor of Andrews, seems to be a variety of this plant, as has been stated by Smith, arising from cultivation. As a proof of this, it may be mentioned that, although sent from East India, no botanist there has been able to detect the same form in the wild state.

XIII. DISCHIDIA. *R. Brown.*—Collyris, *Vahl.*

Corolla urceolata, 5-fida. Corona staminea 5-phylla; foliolis bifidis, lobis patentibus apice recurvis. Antheræ membrana terminatæ. Massæ pollinis erectæ, basi affixæ. Stigma muticum. Folliculi læves. Semina comosa.—Herbæ perennes, in arboribus parasiticæ, lactescentes. Caules ad genicula inferiora radicantes. Folia opposita, crassa, carnosa. Flores parvi.

1. D. *Benghalensis* (Colebr.:) carnosa, caule foliisque lanceolatis glabris, cymis subcapituliformibus breve pedunculatis.—*Colebr. in Linn. soc. trans.* 12. *p.* 357; *Spr. syst.* 1. *p.* 844; *Hook. in bot. mag. t.* 2916; *Wall.! L. n.* 4205. D. lanceolata, *Wall.! Asclep. n.* 62.——Silhet; *Wallich.*—(*R. W.*)

2. D. *Nummularia* (R. Brown:) carnosa, caule foliisque orbicularibus acutis farinosis, pedunculis lateralibus paucifloris.—*R. Brown prod. fl. nov. Holl.* 1. *p.* 461; *Roem. et Schult.* 6. *p.* 57; *Spr. syst.* 1. *p.* 844; *Wall.! L. n.* 4204.—Dischidia, *Wall.! Asclep. n.* 63.—Collyris minor, *Vahl. in act. soc. hist. nat. Hafn.* 6. *p.* 111.—Nummularia lactea minor prima, *Rumph. Amb.* 5. *p.* 472. *t.* 176. *f.* 1.—*Pluk. t.* 354. *f.* 2?— . . . . ; *herb. Finlayson.* Penang et Singapore, etiamque ad ripas Attran in Martabania; *Wallich.*

Vahl's essential character, "foliis oblongo-ovatis," may perhaps belong to a different species, although he refers without doubt to the above figure in Rumphius; with which figure, however, the only specimen I have seen, collected by Finlayson, without any locality specified, accords well.—(*R. W.*)

3. D. *Wallichii* (Wight:) carnosa, caule foliisque orbicularibus glabris haud farinosis.—Dischidia, *Wall.! Asclep. n.* 64.—Nummularia lactea minor secunda, *Rumph. Amb.* 5. *t.* 176. *f.* 2?——Singapore; *Wallich.*

Dr Wallich suggests, in a note attached to the specimens, that this may be Rumphius' Amb. 5. t. 176. f. 1, and the *Collyris minor* of Vahl. I have followed Mr Brown in assigning the former of these synonyms to *D. Nummularia*, but Rumphius' f. 2. of the same plate seems to be the present plant. The specimens I have seen are not in flower.—(*R. W.*)

† 4. D. *cuneifolia* (Wall.)—*Wall. L. n.* 4206.——Attran in Martabania, 1827; *Wallich.*

† 5. D.? *Collyris* (Wall.)—*Wall. L. n.* 4207.—"Collyris major, *Vahl.*—Conchophyllum imbricatum, *Blume.*"——Attran, 1827; *Wallich.*

We have inserted Vahl's description at the end of this memoir: we do not know what work of Blume is referred to.

†6. D. ***Rafflesiana*** (Wall.)—*Wall. L. n.* 4208.——Singapore, 1822; *Wallich.*

Perhaps this may be the same with what is here named *D. Wallichii.*

†7. D. *clavata* (Wall.)—*Wall. L. n.* 4209.——Attran prope Assamee; *Wallich.*

XIV. GYMNEMA. *R. Brown.*

Corolla suburceolata, 5-fida, fauce squamulis denticulisve 5 sinubus insertis nunc coronata. Corona staminea nulla. Antheræ membrana terminatæ. Massæ pollinis erectæ, basi affixæ. Folliculi læves. Semina sæpe marginata, ad umbilicum comosa.—Frutices vel suffrutices sæpius volubiles. Folia opposita, membranacea plana. Umbellæ interpetiolares, cymiformes.

The stamens are not truly naked: in every species we have examined there is a gland-like body, or fleshy leaf on the base of the filaments; in our last section this is so conspicuous as to leave it doubtful whether the species referred to it ought to be retained in the genus, the more especially as this difference in structure is accompanied with a difference in the form of the corolla and general habits of the plants.

§ 1. *Corollæ tubus intus pilosus, fauce denticulis squamulisve 5 decurrentibus coronata.*

1. G. *lactiferum* (R. Brown:) "erectum" vel subvolubile, glabrum, foliis breve petiolatis ovatis obtuse acuminatis sæpe inæqulateris, umbellis petiolo brevioribus plurifloris, corollæ parvæ fauce tuberculis carnosis coronata, tubo intus linea pilosa duplici e tuberculis decurrente instructo.—*Br.! in Wern. soc. mem.* I. *p.* 33; *Roem. et Sch.* 6. *p.* 57; *Spr. syst.* 1. *p.* 844—Asclepias lactifera, *Linn.! sp. p.* 314.——In Zeylana; *Hermann.*

This is so very closely allied to some of the forms of *G. sylvestre,* that I almost doubt if it ought to be kept separate: it is indeed probable that, if we were acquainted with more specimens, so as to exhibit its mode of variation, we should be enabled to prove that it shews a transition from glabrous to pubescent, as *G. sylvestre* certainly does from hairy to nearly glabrous: in which case no practical difference could be pointed out. Unfortunately no specimen is known to exist except those in Hermann's herbarium, which, in succession, Linnæus, Mr Brown, and I myself have described.—(*R. W.*)

2. G. *sylvestre* (R. Brown:) volubile totum foliorum pagina superiore glabra excepta molliter pubescens, foliis ex ovalibus basi apice attenuatis in ovata cordatave acuta, pedunculis petiolum æquantibus, umbellis geminis subcapitatis multifloris parvifloris, stigmate obtuse conico staminibus multo longiore, folliculis tenuibus attenuatis glabris.—*Br. in Wern. soc. mem.* 1. *p.* 33; *Roem. et Sch. syst. veg.* 6. *p.* 57; *Spr. syst.* 1. *p.* 844; *Wall.! Asclep. n.* 71; *Wight! cat. n.* 1530.—G. parviflorum, *Wall.! tent. fl. Nep. p.* 50; *Asclep. n.* 72.—Periploca sylvestris, *Willd. sp.* 1. *p.* 1252; *Phyt.* 1. *p.* 7. *t.* 5. *f.* 3. —Asclepias geminata, *Roxb. fl. Ind.* 2. *p.* 45; *in cœt. merc. Ind. or. mus. tab.* 617.——In Peninsulæ dumetis frequens. Canton in China; *Reeves; Beechey; Millett; Vachell.*

There is certainly no difference between the Chinese (*G. parviflorum,* Wall.) and the Indian plant.

3. G. *hirsutum* (W. & A.:) volubile valde ramosum, ramulis junioribus foliisque subtus hirsuto-tomentosis, foliis supra pubescentibus ex late-ovatis in anguste-oblonga acuminatis, pedunculis petiolo brevioribus, umbellis subcapitatis parvifloris.—*Wight.! cat. n.* 1531.—G. hirsutum? *herb. Wight. olim,* et in *Wall.! Asclep. n.* 70.——Ad Neelgherry; *Wight.*

Perhaps this may prove eventually an alpine variety of the first species. *G. hirsutum,* Wall. tent. fl. Nep., being now ascertained to be a species of *Tylophora,* we have adopted the name given formerly by Dr Wight, although on the erroneous supposition of its being Dr Wallich's plant. *G. attenuatum*, Wall., appears likewise to belong to *Tylophora*.

4. G. *molle* (Wall.:) volubile, ramulis junioribus molliter pubescentibus, foliis e late cordatis in ovata acuminata subtus mollissime pubescentibus supra glabriusculis, umbellis capitatis subsessilibus parvifloris, stigmate truncato stamina vix superante, folliculis brevibus e basi lata sensim ad apicem attenuatis velutinis.—*Wall.! Asclep. n.* 72. *b*, 145, 151.——Ad ripam fluminis Irawaddi haud procul a fontibus petroleo uberibus; *Wallich*.

The character of the fruit is taken from Wall. Asclep. n. 145 and 151; of the flowers from 72, *b*.—(*R. W.*)

4. G. *acuminatum* (Wall.:) volubile, caulis cortice suberosa, ramulis junioribus subtomentosis, foliis ovali-oblongis acuminatis supra glabris subtus tomentosis demum glabrescentibus, pedunculis petiolo brevioribus, umbellis subcapitatis parvifloris, corollæ faucis squamulis exsertis segmentis dimidio brevioribus, folliculis e basi crassa conicis brevibus ligneo-coriaceis demum glabris (2 uncias longis 6–8 lineas basi crassis), seminibus (magnis) ala membranacea marginatis.—*Wall. tent. fl. Nep. p.* 50; *Wall.! Asclep.* 65, 69.—Asclepias acuminata, *Roxb. fl. Ind.* 2. *p.* 55.—Cynanchum acuminatum, *herb. Ham.!*——Chittagong, Silhet, Tavoy, et Maulmyne; *Wallich*.

The Tavoy specimens are more glabrous than the others.—(*R. W.*)

5. G. *latifolium* (Wall.:) volubile?, ramulis junioribus dense pubescentibus, foliis late ovalibus acuminatis utrinque sparsim pilosis parallelim venosis (4–6 uncias longis 3 latis), umbellis petiolo brevioribus cymiformibus multifloris parvifloris, corollæ faucis squamulis inclusis, folliculis subtenuibus acuminatis verrucosis (3 uncias longis 2 lineas basi crassis).—*Wall.! Asclep. n.* 68, 144; *Wight.! cat. n.* 1532.——Tavoy, Silhet, Saluen, Peninsula (fide spec. *herb. Madr.*)

6. G. *glabrum* (Wight:) volubile glabrum, foliis ovatis cordatisve acuminatis parallelim venosis, folliculis magnis oblongo-lanceolatis acuminatis ligneo-coriaceis glabris (4 uncias longis 6 lineas medio crassis).—*Wall.! Asclep. n.* 150 (absque nom. aut hab.)

Very nearly allied to the former, but quite distinct. The specimens are so imperfect that I cannot give a fuller character.—(*R. W.*)

§ 2. *Corollæ faux lineis 5 pilosis decurrentibus.*

The plants of this section have larger flowers than those of the last, and owing to the pedicels being longer, the umbels are larger though fewer-flowered: their appearance is altogether more elegant.

7. G. *tingens* (W. & A.:) volubile frutescens, ramulis junioribus herbaceis glabris, foliis ex cordato-ovatis in ovalia subiter attenuatis utrinque glabris (3–6 uncias longis 2–3 latis), umbellis corymbisve sæpe geminis primum petiolo brevioribus demum spiraliter elongatis, floribus majusculis (pallide flavis), corollæ fauce nuda tubo lineas 5 pilosis instructo, filamentorum glandulis staminibus dimidio brevioribus, stigmate mutico stamina superante.—*Wight.! cat. n.* 1533.—Asclepias montana, *Roxb.* (fol. ovali oblongis) *fl. Ind.* 2. *p.* 45; *in cœt. merc. Ind. or. mus. tab.* 618.—A. tingens, *Ham. in Roxb.* (fol. cordatis) *fl. Ind.* 2. *p.* 53; *Cor.* 3. *p.* 34. *t.* 239; *in cœt. merc. Ind. or. mus. tab.* 1226.—Periploca montana, *Rottl. in herb. Wight.*—Microloma Bidaria, *herb. Ham.! n.* 754; *Wall.! Asclep. n.* 66.—Tylophora Ariadnes, *herb. Ham. n.* 761; *Wall.! Asclep. n.* 54.—*Rumph. Amb.* 5. 40. *f.* 2.?——In sylvis Camrupæ orientalis; *Hamilton*. In montibus Magadhæ; *Hamilton*. Sukanagheer; *Hamilton*. Ad locum Khuriepass dictum; *Royle*.

This plant varies considerably in the form of the leaves, but the spiral tendency of the corymbs or racemes is constant. Hamilton, from the name attached to some of his specimens, seems to have considered this to be the *Co-*

rona Ariadnes, Rumph. Amb. 5. t. 172, but that figure represents rather a species of *Hoya.*

7. G. *elegans* (W. & A.:) volubile, ramis tenuibus glabris lignosis senioribus subverrucosis, foliis cordatis ovatis ovalibusve acuminatis margine undulatis, umbellis breviter pedunculatis demum non spiraliter elongatis pedicellisque æquilongis, floribus parvis (albidis), corollæ fauce nuda tubo intus lineis 5 pilosis instructo, stigmate obtuso staminibus longiore, folliculis sæpius solitariis acuminatis (albidis 3 uncias longis 3–4 lineas crassis), seminibus marginatis.—*Wight.! cat. n.* 1534; *Wall.! Asclep. n.* 67, 155 (absque nom.). ——In montibus Dindygul, Salem, et Columala; *Wight.* Segaen; *Wallich.*

This, though a very different looking plant from the last, and certainly a distinct species, is very difficult to be distinguished by characters. The slender form of every part, and the small few-flowered umbels, present the most striking points of difference.

§ 3. *Corolla ventricosa, fauce edentula, tubo intus lineis pilosis decurrentibus nullis. Filamenta sæpius tuberculo carnoso ad basin instructa.*

The plants referred to this section ought perhaps, as already suggested, to form a distinct genus: they differ widely from the others in general appearance, but more especially in their inflorescence, which in four species forms large lax compound corymbs; in the fifth, however, it returns to the simple umbel. The pollen masses also differ in form from the rest of the genus in being compressed and spathulate, and the appendiculi or corpuscular appendages are thickened and sometimes very long.

8. G. *Nepalense* (Wall.:) foliis ovalibus ovato-oblongisve acuminatis glabris subtus glaucis atque venis fuscis minutim reticulatis, pedunculo communi foliis nunc breviore nunc longiore, umbellis compositis multifloris, floribus breve pedicellatis subcapitatis.—*Wall.! tent. fl. Nep. p.* 50. *t.* 38; *Asclep. n.* 115. *b.*—G. columnare, *Wall.! Asclep. n.* 73. *a.*——Nepal; *Wallich.*—(*R. W.*)

9. G. *columnare* (Wall.:) foliis e basi subcordata lineari-oblongis acuminatis glabris subtus pallide glaucis venisque fuscis reticulatis, umbellis compositis folium subæquantibus multifloris, floribus subsessilibus.—*Wall.! Asclep. n.* 73. *b.*

The only specimens we have seen are from the Calcutta Botanic Garden. This plant is so strikingly like the figure of *G. Nepalense,* that I suspect a specimen of the latter has been figured while one of the other is described. Both species are probably found in Nepal, and perhaps are varieties of the same plant; they differ chiefly in the shape of the leaves.—(*R. W.*)

10. G. *sagittatum* (Wall.:) foliis ovalibus oblongo-ovatisve apice attenuatis glabris subtus albido-glaucis venisque concoloribus, pedunculis tenuibus folia subæquantibus, cymis compositis laxis ramis filiformibus, pedicellis secus lineam elevatam spiralem dispositis, floribus parvis sæpe caducis, folliculis (junioribus) tenuibus glabris.—*Wall.! tent. fl. Nep. p.* 50; *Asclep. n.* 114, *et* 53?——In Silhet; *Wallich.*—(*R. W.*)

11. G. *Wallichii* (Wight:) foliis ovalibus breviter acuminatis glabris subtus pallidioribus venisque concoloribus, cymis racemiformibus, pedicellis secus lineam elevatam spiralem dispositis, corolla ventricosa intus pilosa, gynostegio foliolis 5 carnosis ad basin munito.—*Wall.! Asclep. n.* 135 (absque nom.). Singapore; *Finlayson; Wallich.*

The large fleshy glands, or leaves, at the base of the filaments, renders this a very doubtful species of *Gymnema;* but the three preceding species, with which it associates in other respects, have them also, although not equally conspicuous.—(*R. W.*)

12. G.? *Finlaysonii* (Wight:) volubile-glabrum, foliis cordato-ovatis subiter acuminatis subtus glaucis venis fuscis reticulatis, umbellis densis multifloris petiolo longioribus, corolla subcampanulata glabra, gynostegio brevi

basi nudo, mass. poll. appendiculis elongatis medio spiraliter tortis. *Wall. Asclep. n.* 143 (innom.)——........... ; *Herb Finlayson.*
Probably from China or Singapore.—(*R. W.*)

XV. SARCOLOBUS. *R. Br.*

Corolla rotata, 5-partita. Gynostegium subglobosum. Corona staminea nulla. Antheræ membrana terminatæ. Massæ pollinis erectæ, basi affixæ. Stigma muticum. Folliculi ventricosi carnosi vel coriacei, abortu subsolitarii. Semina marginata scabra.—Frutices volubiles glabri. Folia opposita, glabra, firma. Umbellæ vel corymbi interpetiolares.

We have not been able to examine any species of this genus. It seems to be closely allied to *Gymnema;* nor can we point out any distinction between them except in the seeds being here without a coma, and in the rotate corolla; but indeed Mr Brown says of *Sarcolobus* only, "corolla subrotata." Mr Brown adds two other characters, "faux nuda," and "folliculi carnosi," while in *Gymnema* the faux is "sæpe coronata," and the folliculi "graciles:" as to the first of these, Dr Wallich, although he has detailed the generic character at great length in the 12th vol. of the Asiat. Researches, p. 577, takes no notice of it, leaving it doubtful if applicable to the Bengal species: as to the last, we have seen the folliculi of some species of *Gymnema* very thick, short, and coriaceous, or almost fleshy, as in *Sarcolobus*, but with the seeds comose. Not having at present access to the volume of the Asiatic Researches alluded to, we have derived the characters of Dr Wallich's two species from a comparison of those in Roemer and Schultes's systema veg., and Sprengel's syst.

1. S. *Banksii* (Roem et Sch.:) foliis latiusculis, umbellis multifloris, corolla subrotata fauce nuda, folliculis carnosis.—*Roem. et Sch. syst.* 6. *p.* 58.——Java; in Princes' Island, prope Bataviam; *Banks.*

2. S. *globosus* (Wall.:) foliis ovato-oblongis, cymis bifidis, corolla rotata tubo brevissimo intus villosa, folliculis magnis carnosis globosis utrinque retusis muricatis.—*Wall. in Asiat. res.* 12. *p.* 577, *t.* 4; *List. n.* 4468.——Ad littora subsalsa fluminis Hoogly Benghalæ Australis; Rangoon; et ad ripas flum. Kaluen; *Wallich.*

3. S. *carinatus* (Wall.:) foliis ovalibus oblongisve subcarnosis, pedunculis paucifloris, corolla rotata tubo brevissimo intus glabra, folliculis oblongis acutis lævibus subtus carinatis.—*Wall. in Asiat. res.* 12. *p.* 578. *t.* 5; *List. n.* 4467.—Rangoon; *Wallich.* E. Sunderbun; *Wallich; Hamilton.*

XVI. LEPTADENIA. *R. Brown.*

Corolla subrotata; tubo brevi, fauce squamulis 5 segmentis alternantibus coronata: æstivatio valvata. Corona staminea nulla. Antheræ liberæ, apice simplices. Massæ pollinis erectæ basi affixæ, apice coarctato pellucido. Stigma muticum vel rarius apiculatum. Folliculi læves. Semina comosa.—Frutices volubiles, sæpe cinereo-puberuli. Folia plana. Umbellæ interpetiolares. Stigmatis corpuscula minuta.

1. L. *reticulata* (W. & A.:) volubilis, caulis ramorum seniorum cortice suberoso glabro, ramulis junioribus cinereo-puberulis nunc subtomentosis, foliis ovatis lanceolatisve acutis sæpius glabriusculis nunc breviter albido-pubescentibus, umbellis lateralibus multifloris petiolum subæquantibus, corollæ segmentis margine revolutis intus barbatis, faucis squamulis simplicibus brevibus, stigmate mutico, folliculis subcylindraceo-oblongis obtusis sæpius abortu solitariis.—*a*; foliis breviter farinoso-pubescentibus demum subglabris, pedunculis petiolum subæquantibus pedicellisque æquilongis.—*Wight.! cat. n.* 1535.—Cynanchum reticulatum, *Retz. obs.* 2. *p.* 15; *Willd. sp. pl.* 1. *p.* 1258; *Roem. et Sch. syst. veg.* 6. *p.* 107; *Spr. syst.* 1. *p.* 852.—C. asthmaticum, *herb.*

Ham.! *n.* 766.—Asclepias suberosa, *Roxb. fl. Ind.* 2. *p.* 38; *in cat. merc. Ind. or. mus. tab.* 614.—A. volubilis, *herb. Madr.!*; *Ham.!*; *Wall.! Asclep. n.* 7. (excl. *e.*)—Secamone canescens, *Smith! in Rees' cycl.*—β; ramulis tomentosis, foliis oblongo-lanceolatis, pedunculis brevissimis. — Leptadenia brevipes, *Wight. cat. n.* 1536.—γ; foliis late ovatis basi obtusis retusisve subtus subvelutinis.—Gymnema aurantiaca, *Wall.! Asclep. n.* 147.——α et β; in Peninsula, præsertim in provinciis australioribus. In dumetis Indiæ intra et extra Gangem; *Hamilton.* γ; Prope Pegu in Ava haud procul a fontibus petroleo uberibus; *Wallich.*

Although we have modified the character to contain our second and third varieties, it is possible that both of them may form distinct species. Our α, the type of the species, is a twining shrub common in hedges near the sea coast, and very abundant about Negapatam, where it grows to a great size. Of β we have only a few specimens, mixed with the other; they present young fruit but no flowers; the branchlets are tomentose; the peduncles are very short, not a line and a half long (while in the true plant they are usually as long or even longer than the petiole), and the pedicels are two or three times as long as the peduncles. The var. γ principally differs by the under side of the leaves being almost velvety; of this also we have not seen the flower.

2. L. *imberbis* (Wight:) volubilis glabra, foliis late ovatis cordatisve acuminatis, umbellis lateralibus multifloris petiolum subæquantibus, corollæ segmentis margine revolutis intus imberbibus fauce squamulis simplicibus, stigmate apiculato.—*Wall.! Asclep. n.* 77, *e* (innom.)——Meerut, Suroy; *Wallich.*

The only certain character to distinguish this from the preceding is, that the segments of the corolla are here beardless. (*R. W.*)

3. L. *spartium* (Wight:) suffruticosa glabra, caulibus tenuibus valde ramosis, foliis anguste linearibus apice attenuatis, umbellis subsessilibus, corollæ segmentis intus imberbibus margine revolutis, fauce squamulis latis exsertis, stigmate mutico.—Gymnema spartium, *Wall. Asclep. n.* 76.—Cannychum angustifolium, *herb. Ham. n.* 755.——Ad ripas fluminis Yamuna; *Hamilton.* Himalaiah; *Royle.* Simla; *Dom. Comitissa de Dalhousie.*—(*R. W.*)

XVII. ORTHANTHERA. *Wight.*

Corolla urceolata, 5-fida, fauce nuda, tubo subventricoso: æstivatio valvata. Corona staminea nulla. Antheræ apice simplices, erectæ, simplices, acutæ. Massæ pollinis erectæ, basi affixæ, apice coarctato pellucido. Stigma apiculatum.—Frutex aphyllus, umbellis breve pedunculatis paucifloris. Calyx 5-partitus, segmentis apice subulatis. Corolla extus villosa, intus glabra, gynostegio duplo longior. Filamenta basi glandula aut foliolo carnoso basi, ut in *Gymnadenia,* munita. Antheræ liberæ, stigmati apiculato incumbentes. Folliculi ignoti.

This is perhaps too nearly allied to *Leptadenia,* from which it only differs by the urceolate corolla, the absence of scales to the orifice of the tube, and erect anthers; but being unwilling to propose a farther modification of the character of that genus, I have considered it safer to constitute the present one. The habit is very different from that of *Leptadenia.*

1. O. *viminea* (Wight).—Apocynea viminea, *Wall.! cat. n.* 7478.——Doab Hindusthaniæ; *Royle,* 1825.—(*R. W.*)

XVIII. TYLOPHORA. *R. Brown.*

Corolla rotata, 5-partita. Corona staminea 5-phylla; foliolis simplicibus, carnosis. Antheræ membrana terminatæ. Massæ pollinis transversæ vel subascendentes, minutæ, ventricosæ. Stigma muticum. Folliculi læves.

apice attenuati, compressi, hinc subangulati. Semina comosa.—Herbæ aut frutices volubiles. Umbellæ interpetiolares, solitariæ vel alternatim secus pedunculum longum flexuosum dispositi. Flores sæpius parvi.

The extreme minuteness of the pollen-masses renders it difficult to determine the situation of this genus. The form, however, of the gynostegium and leaves of the corona, readily distinguish it from every other except *Iphisia*, but which differs by the large size of the decidedly pendulous pollen-masses. In examining the pollen of *Tylophora*, considerable care is required, as the masses, when removed from their cells, sometimes appear erect, sometimes pendulous, in the same species, nay even in the same flower: this is so much the case in *T. asthmatica*, that it was long before we could decide upon actually referring it to the genus. As now defined, we believe this genus to be quite distinct from *Hoya*, to which Mr Brown, when he first established it, considered it too closely allied.

§ 1. *Cor. st. foliola obtusissima vel obtuse acuminata intus ecuspidata gynostegio breviora.*

1. T. *carnosa* (Wall.:) volubilis glabra, caulibus ramisque tenuibus, foliis carnosis ovatis subcordatisve mucronatis nitidis subtus pallidis (1–2 uncias longis 6–9 lineas latis), pedunculis flexuosis ad flexuras pedicellos complures filiformes gerentibus, floribus parvis (intus purpureis), cor. st. foliolis carnosis suborbicularibus, massis pollinis adscendentibus, stigmate convexo, folliculis glabris sæpius abortu solitariis.—*Wall.! Asclep. n.* 49; *Wight.! cat. n.* 1537.—Asclepias dichotoma, *herb. Rottl.!*——In Amherst; *Wallich.* Ad Colombo in Ceylona; *Rottler.* In arenosis prov. Tanjorensis; *Wight.*
A species apparently undescribed.

2. T. *tenuissima* (W. & A.: volubilis glabra, ramis tenuibus, foliis oblongo-lanceolatis basi subcordatis cuspidatis aveniis margine subrevolutis (2–3 uncias longis 3–6 lineas latis), pedunculis flexuosis ad flexuras pedicellos complures filiformes gerentibus, floribus parvis, cor. st. foliolis ovato-oblongis, massis poll. adscendentibus, stigmate convexo, folliculis divergentibus glabris.—*Wight.! cat. n.* 1538.—Asclepias tenuissima, *Roxb. fl. Ind.* 2. *p.* 41; *in cat. merc. Ind. or. mus. tab.* 1383; *Roem. et Sch. syst. veg.* 6. *p.* 85.——In Peninsula; *Wight.* In Benghala; *Roxburgh.*
Very nearly allied to the last, and perhaps only a variety. The leaves, however, are nearly twice the length, and much narrower in proportion.

3. T. *pauciflora* (W. & A.:) volubilis glabra, ramis tenuibus, foliis longiuscule petiolatis e basi lata ovata vel subcordata sensim attenuatis margine subundulatis, pedunculis folio brevioribus flexuosis pedicellos binos ternosve filiformes ad flexuras gerentibus, floribus parvis, calycis corollæque segmentis obtusis, cor. st. foliolis late-ellipticis obtusis, massis poll. adscendentibus, stigmate convexo, folliculis glabris.—*Wight.! cat. n.* 1539.——Apud Courtallum; *Wight.* In Benghalæ provinciis septentrionalibus; *Royle.*
This species is easily distinguished from all the others by the almost triangular form of the leaves.

4. T. *hirsuta* (Wight:) volubilis tota pubescens, foliis late ovatis acuminatis (3–5 uncias longis 2–3 latis), pedunculis folium subæquantibus flexuosis ad flexuras umbellam sessilem gerentibus, pedicellis filiformibus, floribus parvis, corollæ segmentis ovatis obtusis undulatis, cor. st. foliolis ovatis sub-acuminatis obtusis, massis poll. transversis, stigmate obtuso, folliculis tenuibus.—Gymnema hirsutum, *Wall.! tent. fl. nep. p.* 50; *Asclep. n.* 46.——Apud Chandagery et Sanko; *Wallich.*—(*R. W.*)

5. T. *mollissima* (Wall.:) volubilis tota pubescens, foliis ovalibus vel cordato-ovatis acuminatis mucronatis (1–2 uncias longis 1–1¼ latis) pedunculis folio duplo triplove longioribus flexuosis ad flexuras umbellam sessilem gerentibus, pedicellis filiformibus, floribus parvis, corollæ segmentis ovatis obtusis undulatis, cor. st. foliolis apice truncatis vel rotundatis, massis poll. trans-

versis, stigmate obtuso, folliculis pubescentibus.—*Wall.! Asclep. n.* 47; *Wight.! cat. n.* 1540.——Ad Neelgherry et in montibus Dindygul; *Wight*................ *Herb. Finlayson.*

Perhaps too nearly allied to the preceding, from which it seems principally to differ in the size of the leaves.

6. T. *rotundifolia* (Ham.:) volubilis tota breviter pubescens, foliis rotundis vel rotundato-ovatis (circiter uncias duas latis), pedunculis folio brevioribus pedicellos subæque longos filiformes in fasciculos subduos dispositos prope apicem gerentibus, floribus majusculis glabris, cor. st. foliolis carnosis brevibus medio gynostegio adnatis, massis poll. transversis, stigmate obtuso.—*Herb. Ham.! n.* 759; *Wall.! Asclep. n.* 120.——In montibus et campis elevatis Camrupæ orientalis; apud Dongtala et Habora; *Hamilton.*—(*R. W.*)

7. T. *fasciculata* (Ham.:) erecta aut subvolubilis glabra, foliis approximatis ovatis obtusiusculis subcarnosis (1–2 uncias longis et unam fere latis) versus ramorum apices decrescentibus, pedunculis erectis flexuosis ad flexuras fasciculos 2–3 paucifloros gerentibus, cor. st. foliolis oblongo-ovatis obtusiusculis, massis poll. transversis, stigmate apiculato.—*α, parviflora;* caulibus apice subvolubilibus, floribus minutis.—T. fasciculata, *Herb. Ham.! n.* 758; *Wall.! Asclep. n.* 55.—*β, macrantha;* radice fasciculato, caulibus lignosis erectis, fasciculis approximatis, floribus majusculis, corollæ segmentis ligulatis.—*Wight.! cat. n.* 1541; *Wall.! Asclep. n.* 31 (innom.)——*α*; In Sylvis Mithilæ, apud Sukanagur; *Hamilton*—*β*; Ad Neelgherry; *Wight.*

In this species the corpuscles of the stigma differ from those of all the other *Asclepiadeæ* which we have examined, in so far as the two portions, into which Brown has observed that it is divisible, are here herbaceous and very slightly united, and the extremities at both ends divaricate from each other. The pollen masses are perfectly transverse, and are attached by a very slender appendiculus. Perhaps it might, from this peculiarity, form a distinct genus, in which case the two varieties may prove species; but our specimens of both are too few in number to permit of our deciding what distinctive characters may prove constant between them.

8. T. *longifolia* (Wight:) volubilis glabra, foliis oblongo-lanceolatis acutis basi subcordatis (6–8 uncias longis 1–1½ latis), paniculis folio brevioribus multifloris, ramis flexuosis ad flexuras umbellam parvam gerentibus, corollæ segmentis obtusis, cor. st. foliolis gynostegio subdimidio brevioribus late ellipticis obtusis, massis poll. transversis, stigmate convexo.—*Wall.! Asclep. n.* 48. (innom.)——In Silhet; *Wallich.*—(*R. W.*)

9. T. *exilis* (Colebr.:) volubilis glabra, foliis ovato-lanceolatis acuminatis (2–3 uncias longis unam latis), paniculis majusculis folio duplo triplove longioribus multifloris, ramis flexuosis ad flexuras pedicellos umbellatos gerentibus, corollæ segmentis acutiusculis, cor. st. foliolis late ellipticis obtusissimis gynostegio brevioribus, massis poll. transversis, stigmate apiculato.—*Colebr. in Linn. soc. trans.* 12.—Pergularia exilis, *Spr. syst. veg.* 1. *p.* 844; *Wall.! Asclep. n.* 51 (innom.)——Silhet; *Wallich.*

Very nearly allied to the last, but certainly distinct.—(*R. W.*)

§ 2. *Cor. st. foliola submembranacea tenuiter acuminata acuta intus ecuspidata gynostegio longiora.*

10. T. *tenerrima* (Wight:) volubilis tenella glabriuscula, foliis anguste lanceolatis apice attenuatis (1–2 uncias longis 1–4 lineas latis), pedunculis filiformibus folio multo longioribus florum duos tresve fasciculos distantes gerentibus, corolla alte 5-fida, cor. st. foliolis gynostegium superantibus ovatis apice tenuiter acuminatis, massis poll. transversis, stigmate obtuso.—Homo-

lostyles tenerrima, *Wall.! Asclep. n.* 96, 118.——Ad Purpuhna, in Nepala; *Wallich.* In montibus Himalaiah, alt. 1000 hexapod.; *Royle.*—(*R. W.*)

§ 3. *Cor. st. foliola depressa carnosa gynostegii basin amplectentia, margine interiore apice in denticulum cuspidiformem erectum producto.*

11. T. *capparidifolia* (W. & A.:) volubilis glabra, ramis tenuibus, foliis elliptico-oblongis mucronato-acuminatis coriaceis reticulatim venosis (1½–2 uncias longis 5–7 lineas latis), racemis umbelliformibus subsessilibus paucifloris parvifloris, corollæ segmentis obtusis glabris, cor. st. foliolis rotundatis carnosis apice intus denticulo brevi acuto auctis, mass. poll. subadscendentibus, stigmate obtuso.—*Wight.! cat. n.* 1542.—Asclepias tenuis, *herb. Heyn.!; Wall.! Asclep. n.* 122.——In Mysore? *Heyne.*

The leaves of this species are remarkable for their resemblance to those of some species of *Capparis*, particularly to the narrow-leaved forms of *C. stylosa* and *heterophylla.* The name given by Heyne, although applicable to it, on the supposition of its being an *Asclepias*, is not so as a *Tylophora.*

12. T. *purpurea* (Wall.:) volubilis glabra, ramis tenuibus, foliis subcordato-oblongis mucronato-acuminatis (1–1½ uncias longis 8–9 lineas latis), pedunculis filiformibus folium subæquantibus florum fasciculos 2–3 parvos gerentibus, corollæ segmentis acutis, cor. st. foliolis rotundatis gynostegii basin amplectentibus apice intus denticulo brevi auctis, massis poll. transversis, stigmate obtuso.—*Wall.! Asclep. n.* 50.——Ad ripas fluminis Irawaddi; *Wallich.*—(*R. W.*)

13. T. *asthmatica* (W. & A.:) volubilis glabra aut pubescens, ramis tenuibus, foliis ovatis subrotundisve acuminatis basi sæpius cordatis supra glabris, pedunculis folio brevioribus umbellas 2–3 sessiles paucifloras versus apicem gerentibus, floribus majusculis longe pedicellatis (purpurascentibus), corollæ segmentis acutis, cor. st. foliolis carnosis depressis gynostegii basin amplectentibus apice intus dente elongato erecto auctis, massis poll. tranversis minutis, stigmate obtuso, folliculis glabris divaricatis.—Asclepias asthmatica, *Roxb. fl. Ind.* 2. *p.* 33; *Willd. sp. p.* 1270; *Roem. et Sch. syst. veg.* 6. *p.* 80; *Spr. syst.* 1. *p.* 849; *Wall.! Asclep. n.* 116. *b.*—A. tunicata, *Wall.! Asclep. n.* 117. *c.*—A. vomitoria, *Koen. in herb. Banks; Roxb. in cæt. merc. Ind. or. mus. tab.* 608.—Cynanchum Ipecacuana, *Willd.*—C. vomitorium, *Lam. enc. meth.* 2. *p.* 235.—C. viridiflorum, *Sim's bot. mag. t.* 1929; *Spr. syst.* 1. *p.* 852.—Hoya planiflora, *Wall.! Asclep. n.* 117. *b.*—Tylophora pubescens, *Wall.! Asclep. n.* 117. *a.*—*Wall.! Ascl. n.* 119, 121, 148 (absque nom.)—*Pluk. t.* 336. *f.* 7.—*α, pubescens;* ubique adpresse et sæpe molliter pubescens foliis supra exceptis.—*Wight.! cat. n.* 1543.—*β, glabra;* tota glabra.—*Wight. cat. n.* 1543. *b.*——In Peninsula; *Roxburgh; Klein; Wight.* In colle contra Loongye; *Wallich.* Ad ripas prope Yenangheun; *Wallich.* Ad sinum Siamicum; *Finlayson.* Ad Segaen; *Wallich*, 1826.

This has, if we mistake not, been described and figured in some foreign work as *Hoya viridiflora.* It is probable that *Cynanchum tomentosum*, Lam., ought also to be referred here; but the description is very imperfect. We can see no other difference between the two varieties than what we have mentioned: there are even some intermediate states as to pubescence, such as Wall. Ascl. n. 117, *a* and *c*, which exhibit only a sprinkled pubescence.—It was not till after repeated examinations that we satisfied ourselves of the position of the pollen masses of this plant, and that they agreed with several undoubted Indian species of the genus. At first they and the appendiculi are horizontal; but immediately on being removed from the anther-cases, they are sometimes erect from the extremity of the horizontal appendiculi, and sometimes appear pendulous from the appendiculi being flexuose.—We have to regret the want of specimens of the New Holland species, from which circum-

stance we may have so altered the generic character as to render it less applicable to them than was contemplated by the founder of the genus.

XIX. BELOSTEMMA. *Wall. ined.*

Corolla rotata. Corona staminea 5-phylla; foliolis lineari-oblongis carnosis gynostegii medio verticaliter affixis aversis patentibus. Antheræ membrana terminatæ. Massæ pollinis globosæ, subadscendentes vel transversim sub stigmatis margine dilatato sitæ. Stigma muticum.—Suffrutex volubilis ubique laxiuscule pubescens. Folia opposita, cordato-ovata, subacuminata. Pedunculi folio breviores. Umbellæ simplices, pauciforæ. Flores parvi.

This genus differs from *Tylophora* only in the form of the corona. In that, the leaflets are close-pressed on the gynostegium; in this they appear as if attached by the one end, and stand at right angles with the gynostegium, and transversely across the corolla.

1. B. *hirsutum* (Wall.)—*Wall.! Asclep. n.* 153.——In Nepala; *Wallich*.—(*R. W.*)

XX. IPHISIA. *W. & A.*

Corolla rotata, 5-fida. Corona staminea 5-phylla; foliolis carnosis, versus apicem subiter angustatis. Antheræ membrana terminatæ. Massæ pollinis ventricosæ, apice affixæ, ex appendiculi adscendentis filiformis extremitate pendulæ. Stigma muticum. Folliculi læves apice attenuati, glabri. Semina comosa.—Herbæ vel suffrutices suberecti aut volubiles. Folia opposita. Flores parvi. Umbellæ interpetiolares, simplices, aggregatæ vel ad flexuras pedunculi subsessiles.

This is very nearly related to *Tylophora* both in habit and structure of the flowers, but is separated on account of the truly pendulous pollen masses.

1. I. *multiflora* (W. & A.:) suffruticosa volubilis glabra, foliis ovalibus cuspidato-acuminulatis petiolatis, umbellis sessilibus ad flexuras pedunculi communis alternatim dispositis, corollæ segmentis brevibus, appendiculis elongatis basi sursum apice deorsum flexis.—*Wight.! cat. n.* 1544.——Ad Neelgherry; *Wight.*

2. I. *Govanii* (Wight:) suberecta, foliis subsessilibus e late ovalibus breviter acuminatis in lanceolata acuta, umbellis simplicibus aggregatis paucifloris, pedicellis pedunculos subæquantibus, corollæ segmentis lanceolatis intus glabris extus hirsutis, appendiculis breviusculis primo suberectis.—*Wall.! Asclep. n.* 131 (absque nom.)——In montibus Himalaiah; *Govan; Royle.*

If the retinacula be examined before the pollen masses are removed from the anther-cases, they will be found erect with a very slight curvature both at the base and apex; but as soon as they are removed from their natural situation, they become almost horizontal and nearly straight.

XXI. PENTATROPIS. *R. Brown.*

Corolla rotata, 5-fida. Corona staminea 5-phylla; foliolis antheris oppositis, gynostegio verticaliter adnatis, aversis, apice solutis. Massæ pollinis ventricosæ, pendulæ, infra apicem affixæ. Stigma muticum. Folliculi læves. Semina comosa.—Suffrutices volubiles. Folia subcarnosa, plana. Umbellæ interpetiolares, pauciforæ, subsessiles. Folliculi obsolete trigoni, hinc planiusculi.

1. P. *microphylla* (W. & A.)—*Wight.! cat. n.* 1545.—Asclepias microphylla, *Roxb. fl. Ind.* 2. *p.* 35; *Heyne in Roth. nov. sp. p.* 177; *Roem. et Sch. syst. veg.*

6. *p.* 85; *Spr. syst.* 1. *p.* 849; *Wall.! Asclep n.* 111.—A. tenuiflora, *Roxb. in cæt. merc. Ind. or. ms. tab.* 610; *herb. Russ.!*—Oxystelma caudata, *herb. Ham.! n.* 771; *Wall.! Asclep. n.* 110.——In Coromandelia, frequens. In Benghala, ad ripas argillosas fluminis " Yamuna" dicti; *Hamilton.*

Asc. alexiaca of Jacquin may be this plant, but the figure he refers to (Rheed. Mal. 9. t. 13), and from which he in part made his description, bears no resemblance to what we have in view: Roxburgh supposes that synonym may belong to *Tylophora asthmatica,* but there are many obvious reasons against that affinity. We can perceive no difference between the Bengal and Coromandel plants. As we are unacquainted with *P. cynanchoides* (Brown in Salt. abyss. app.), or in what respect it differs from the Indian species, we are unable to give a specific character: our species may probably, however, be sufficiently distinguished as follows: Volubilis glabra, foliis subcarnosis ovatis mucronatis basi rotundatis, subcordatisve latitudine $1\frac{1}{4}$–2-plo longioribus (6–18 lineas longis 3–12 latis), umbellis subsessilibus paucifloris, pedicellis elongatis filiformibus, calyce minuto, corolla patenti-reflexa segmentis acutis, cor. st. foliolis latis basi aversis apice cuspidatis incurvis gynostegium æquantibus.

XXII. CALOTROPIS. *R. Brown.*

Corolla subcampanulata, tubo angulato, angulis intus saccatis, limbo 5-partito. Corona staminea 5-phylla; foliolis carinæformibus, gynostegio verticaliter adnatis, basi recurva. Antheræ membrana terminatæ. Massæ pollinis compressæ, apice attenuato affixæ, pendulæ. Stigma muticum. Folliculi ventricosi, læves. Semina comosa.—Frutices erecti, glabri, lactescentes. Folia opposita, lata. Umbellæ interpetiolares. Flores majusculi, speciosi.

§ 1. *Folia subsessilia cuneato-oblonga vel late ovata basi retusa vel subcordata.*

1. C. *gigantea* (R. Brown:) corollæ segmentis patentibus marcescentibus reflexis margine revolutis, cor. st. foliolis gynostegio brevioribus basi obtusa circinnatim recurva apice incurvo subtridentato.—*Brown in hort. Kew.* (*ed.* 2) 2. *p.* 78; *Roem. et Sch. syst. veg.* 6. *p.* 91; *Spr. syst.* 1. *p.* 850: *Ham.! herb. n.* 775; *in Linn. soc. trans.* 14. *p.* 245; *Wall.! Asclep. n.* 93; *Wight.! cat. n.* 1546.—Asclepias gigantea, *Willd. sp.* 1. *p.* 1264; *Roxb. fl. Ind.* 2. *p.* 30; *in cæt. merc. Ind. or. mus. tab.* 606.—*Rheed. Mal.* 2. *t.* 31; *Rumph. Amb.* 7. *t.* 14. *f.* 1.——In Coromandelia, frequens; *Wight.* In Indiæ arenosis vulgatissima; *Roxburgh*; *Hamilton.*

This is the Mador or Mudor plant, from which the late Dr Duncan of Edinburgh obtained that singular substance called Mudarine, which possesses the property of coagulating by heat, and becoming again fluid on exposure to cold.

2. C. *Hamiltonii* (Wight:) corollæ segmentis erectis, cor. st. foliolis gynostegium æquantibus basi acuminata recurva apice bipartito patulo.—C. procera, *Ham.! herb. n.* 777; *in Linn. soc. trans.* 14. *p.* 426 (excl. syn.); *Wall.! Asclep. n.* 92.—Asclepias madaree, *hort. bot. Calc.*——In Methilæ Magadhæ et Cosalæ arenosis; *Hamilton.* Apud Patna; *Hamilton.* Doab Hindustaniæ; *Royle.*

The Persian *C. procera,* judging from Andrews' figure, appears very distinct.—(*R. W.*)

3. C. *Wallichii* (Wight:) corollæ segmentis patulis, cor. st. foliolis gynostegium superantibus basi obtusa recurva apice supra stigma incurvato subbilobo.—C. gigantea, *Wall.! Asclep. n.* 91, *b.*——Melloon; *Wallich.*

In addition to the above marks of distinction, this differs in habit from both the preceding: the flowers are in loose long-peduncled panicles, in place of contracted cymes or umbels.—(*R. W.*)

4. C. *heterophylla* (Wall.:) corollæ segmentis erectiusculis, cor. st. foliolis gynostegium æquantibus basi obtusa recurva apice stigmatis margini incumbente bilobo lobis incrassatis subdivergentibus.—*Wall.! Asclep. n.* 91, *a.*——Ad ripas fluminis Irawaddi; *Wallich.*

The flowers of this are smaller than any other of the Indian species, and form small compact umbels on short peduncles.—(*R. W.*)

§ 2. *Folia petiolata cuneato- vel elliptico-oblonga.*

5. C. *herbacea* (Wight:) umbellis compositis, corollæ tubo globoso gynostegium subincludente laciniis patulis, cor. st. foliolis gynostegio brevioribus basi acuta patenti-recurva apice tridentato gynostegio incumbente.—*C. acia, Ham.! in Linn. soc. trans.* 14. *p.* 247; *Wall.! Asclep. n.* 90.—Asclepias herbacea, *Roxb. fl. Ind.* 2. *p.* 50; *in cæt. merc. Ind. or. mus. tab.* 1544.——In Bengalæ provinciis remotioribus; *Carey.* In Mithulæ borealis sylvis; *Hamilton.*—(*R. W.*)

XXIII. OXYSTELMA. *R. Brown.*

Corolla subrotata, patens tubo brevi. Gynostegium exsertum. Corona staminea 5-phylla, foliolis acutis compressis indivisis. Antheræ membrana terminatæ. Massæ pollinis compressæ, apice attenuato affixæ, pendulæ. Stigma muticum. Folliculi læves. Semina comosa.—Suffrutices volubiles glabri. Folia opposita. Racemi vel umbellæ interpetiolares.

What Kunth describes as an external corona in the allied genus *Philibertia*, is probably only a mere contraction and consequent incrassation of the tube of the corolla, which may likewise be observed in the present genus.

1. O. *esculentum* (R. Brown:) foliis lineari-lanceolatis venosis, corolla margine ciliata, folliculis oblongis acuminatis.—*Roem. et Sch. syst. veg.* 6. *p.* 89; *Spr. syst. p.* 850; *Wall.! Asclep. n.* 94; *Wight.! cat. n.* 1547.—Periploca esculenta, *Linn. fil. suppl. p.* 168; *Willd. sp. pl.* 1. *p.* 1250; *Roxb. Cor.* I. *p.* 13. *t.* 11.—Asclepias rosea, *Roxb. fl. Ind.* 2. *p.* 40.—*Pluk. t.* 359. *f.* 6.——Per omnem Indiam.

2. O. *Wallichii* (Wight:) foliis anguste lineari-lanceolatis, corolla margine ciliata, folliculis breve-ovalibus obtusis.—*Wight.! cat. n.* 1548?—O. esculentum, *Wall.! Asclep. n.* 95.——Ad ripas Gangis atque Irawaddi; *Wallich.* Prope Pathacottah, in prov. Tanjorensi?; *Wight.*

The shape of the follicles at once distinguish this species, so much alike that I can point out no other certain character. In *O. esculentum* they are between two and three inches long, and taper to a point; in *O. Wallichii* they scarcely exceed an inch, and are equally obtuse at both ends, presenting when compressed by drying an oval form. The leaves in *O. Wallichii* are considerably narrower than in most states of *O. esculentum,* on which account I think it probable, in the absence of fruit, that my specimens from Pathacottah may belong to it: for the same reason, beautiful specimens collected "near Kishnagur on the banks of the river Tellinghie," by the Countess of Dalhousie, may be referred rather to this than to *O. esculentum.*—(*R. W.*)

XXIV. RAPHISTEMMA. *Wall.*

Corolla companulata, limbo 5-partito. Corona staminea 5-phylla; foliolis compressis elongatis compressis. Antheræ membrana terminatæ. Massæ pollinis sub apice affixæ, pendulæ. Stigma obtuse conicum. Folliculi sæpius (abortu) solitarii, subventricosi. Semina comosa.—Frutex glabra, volubilis. Folia majuscula, cordata. Flores albidi, corymbosi. Cor. st. foliola exserta, corollæ limbum subæquantia.

This genus, as Dr Wallich suggests, is very nearly allied to *Kanahia*; the difference appearing to depend more on habit and geographical situation than on the characters afforded by the flower.

1. R. *pulchellum* (Wall.)—*Wall. pl. as. rar.* 2. *p.* 50. *t.* 163; *List, p.* 158. *n.* 1310; *Asclep. n.* 108, 109?—Asclepias pulchella, *Roxb. fl. Ind.* 2. *p.* 54.—Pergularia? campanulata, *herb. Ham.*——Silhet; *Smith.* Gualpara; *Hamilton.* Tavoy; *Gomez.* In Pegu prope Rangoon; *Wallich.* Ad sinum Siamicum?; *Finlayson.*

The only specimen I have is from Finlayson's herbarium (Wall. n. 109); it seems to differ a little from Wallich's figure and description, but it is too imperfect to enable me to determine whether it be a distinct species.—(*R. W.*)

XXV. HOLOSTEMMA. *R. Brown.*

Corolla subrotata, 5-fida. Corona staminea imo gynostegio inserta, simplex, annularis, obsolete 5-loba. Antheræ membrana terminatæ. Massæ pollinis pendulæ, compressæ, apice attenuato affixæ. Stigma muticum. Folliculi ventricosi, læves. Semina comosa.—Frutices glabri, volubiles. Folia opposita, lata. Umbellæ interpetiolares, breviter pedunculata. Flores speciosi.

1. H. *Rheedei* (Spr.:) foliis late-ovatis cordatis.—*Spr. syst.* 1. *p.* 851; *Wall.! pl. as. rar.* 2. *p.* 51; *List, n.* 4469; *Wight.! cat. n.* 1549.—H. Ada-Kodien, *Roem. et Sch. syst. veg.* 6. *p.* 95.—Asclepias annularia, *Roxb. fl. Ind.* 2. *p.* 37; *in cæt. merc. Ind. or. mus. tab.* 613.—A. convolvulacea, *herb. Heyne.!*—Cynanchum annulare, *herb. Heyne.!*—C. cordifolium et grandiflorum, *herb. Russ.!*—Sarcostemma annulare, *Roth. nov. sp. p.* 178; *Roem. et Sch. syst. veg.* 6. *p.* 115; *Spr. syst.* 1. *p.* 854.—Gomphocarpus volubilis, *herb. Ham.!*——In Malabaria; *Rheede.* Mysore; *Heyne; Wight.* Samulcottah et in montibus Circars; *Russell; Roxburgh.* Gualpara; *Hamilton.* Kenour; *Royle.*

This species seems to have a very extensive distribution, being found in Malabar and Mysore in the south, in the Circar district in the north of the Peninsula, and apparently the same plant at the foot of the Himalaiah mountains. Between the specimens we have examined from Mysore and those from Simla and Kenour in the Himalaiah range, we can see no specific difference. They are, however, imperfect.

2. H. *fragrans* (Wall.:) "foliis oblongo-cordatis summis oblongo lanceolatis basi cordato-sagittatis lobis subimbricantibus."—*Wall. pl. as. rar.* 2. *p.* 51; *List, n.* 4470.——In montosis ripæ Irawaddi ad Seendya et Prome, etiam in monte Taong Dong, haud procul ab oppido Ava; *Wallich.*

This I have not seen.

XXVI. CYNANCHUM. *Linn.; R. Brown.*

Corolla subrotata, 5-partita. Corona staminea gamophylla, 5–20-loba, dum 5-loba lobis antheris oppositis. Antheræ membrana terminatæ. Massæ pollinis ventricosæ, pendulæ. Stigma sæpius apiculatum, nunc muticum, rarius incluso-rostratum. Folliculi læves. Semina comosa.—Herbæ perennes, vel suffrutices, ut plurimum volubiles. Folia opposita. Umbellæ interpetiolares.

§ 1. *Cor. st. tubulosa, ore 5–10-fido, gynostegium includens; lacinulæ 5 interiores, antheris lobisque exterioribus oppositæ et parallelæ.*

1. C. *Dalhousiæ* (Wight:) volubile glabriusculum, foliis lineari-lanceolatis attenuatis basi latioribus cordatis subtus pallidis venisque fuscis notatis

(4–6 uncias longis, 4–12 lineas basi latis), umbellis paucifloris, pedicellis pedunculo longioribus, corollæ segmentis ovalibus obtusis, cor. st. corollam æquante 10-loba lobis alternis minoribus laciniis interioribus parvis obtusis dentiformibus, stigmate subapiculato indiviso.——Ad Simla; *Dom. Comitissa de Dalhousie*. Mussorie; *Royle*. Augusto mense floret.—(*R. W.*)

§ 2. *Cor. st. tubulosa, ore 5–10-fido, gynostegium includens; carinæ* 10 *interiores simplices vel in lacinulas productæ.*

2. C. *corymbosum* (Wight:) volubile glabrum, foliis cordato-ovatis acuminatis subtus glaucis, corymbis petiolo longioribus multifloris, cor. st. corollam æquantis 10-fidæ lobis alternis minoribus carinis interioribus simplicibus, massis poll. infra apicem affixæ, stigmate apiculato bifido.—Cynanchum, *Wall.! Asclep. n.* 81.——Silhet; *Wallich*.—(*R. W.*)

§ 3. *Cor. st. tubulosa, ore 5–10-fido, gynostegium includens; lacinulæ carinæve interiores nullæ.*

3. C. *pauciflorum* (R. Brown:) volubile glabrum, foliis ovatis acuminatis basi reniformi-cordata auriculis divergentibus, umbellæ paucifloræ, pedunculis petiolo brevioribus, floribus glabris breviter pedicellatis, cor. st. corollam æquantis ore plicato 10-fido intus nudæ lobis antheris oppositis lanceolatis acuminatis apice bifido alternis brevissimis truncatis ac emarginatis, massis poll. infra apicem affixis, stigmate apiculato obtuso vix emarginato.—*R. Br. in Wern. soc. mem.* 1. *p.* 45; *Roem. et Sch. syst. veg.* 6. *p.* 98; *Spr. syst.* 1. *p.* 852; *Wight.! cat. n.* 1550.—Cynanchum, *Wall.! Asclep. n.* 79.—Periploca tunicata, *Retz. obs.* 2. *p.* 15; *Willd. phyt.* 1. *p.* 7. *n.* 23. *t.* 5. *f.* 3; *sp.* 1. *p.* 1252; *herb. Rottl.!* (*ex hort. miss.*)—Asclepias tunicata, *Roxb. fl. Ind.* 2. *p.* 35; *in cæt. merc. Ind. or. mus. tab.* 609.——In sepibus frequens. In arenosis apud Vellanganny prope Negapatam, copiose.

We have presumed to remove this species from the section to which Mr Brown had referred it, as there are no interior keels, but only folds of the corona which arise from the contraction of its mouth and disappear when it is expanded. In every other respect our plant agrees with his character, and also with Willdenow's figure and description. The follicles are ventricose. Perhaps this and *C. angustifolium* ought to be described with only five lobes to the corona, the alternate ones being extremely short; but they are really present, and even when least conspicuous may be inferred from the broad sinus or space that separates the longer ones from each other.

4. C. *Wallichii* (Wight:) volubile glabrum, foliis oblongo-ovatis acuminatis basi cordata auriculata auriculis subincurvis, corymbis folio brevioribus multifloris, pedicellis pedunculum æquantibus vel superantibus, cor. st. corollam æquantis sub-10-fidæ lobis antheris oppositis longioribus bifidis, massis poll. infra apicem affixis, stigmate mutico.—Cynanchum, *Wall.! Asclep. n.* 80. ——Jentaga, 1828, et Chuna Poonjee; *Wallich*.

Very closely allied to *C. pauciflorum*, but I think quite distinct; the leaves deeply cordate with the lobes ear-like and approximating, and the pedicels usually longer than the peduncles, are striking points of difference.—(*R. W.*)

5. C. *Callialata* (Ham.:) volubile glabrum, foliis ovatis vel ovalibus acuminatis basi cordatis sinu angusto subtus glaucis, umbellis petiolo brevioribus, pedicellis pedunculum æquantibus, cor. st. 10-lobæ corolla subduplo brevioris lobis antheris oppositis apice bifidis alternis brevissimis, stigmate subapiculato integro, folliculis alatis.—*Herb. Ham.! n.* 767; *Wall.! Asclep. n.* 83; *Wight.! cat. n.* 1551.—C. auriculatum, *herb. Ham.! n.* 768; *Wall.! Asclep. n.* 85.—Periploca tunicata, *herb. Madr.!*; *Wall.! Asclep. n.* 78, *b, c* (*a* non visa).——In India Gangetica; *Hamilton*. Ad Columala; *Wight*.

Hamilton's *C. auriculatum* presents a different aspect, particularly in the colour of the leaves of the dried specimens, which is pale ochry, and in the shorter peduncles and pedicels, and consequently smaller and more compact umbels: but we can point out no other characters. We believe this has been sometimes confounded with the *Asclep. tunicata* of Koenig, and indeed Hamilton, in his herbarium, asserts that this is what Roxburgh affirmed to him to be that plant: they are closely allied, but easily distinguished by the base of the leaves, in this with an acute sinus, in *C. pauciflorum* with a broad sinus and diverging lobes, and by the relative size of the calyx, corolla, and crown. In this, the calyx equals the crown, and the corolla is almost twice as long; while in the other the corona is almost as long as the corolla, and both of them twice the length of the calyx. We do not know the etymology of the specific name: perhaps the plant is so called in some districts by the natives; if it be in allusion to the winged fruit, it is altogether inadmissible, being a compound of Greek, Latin, and bad grammar: we had perhaps acted more prudently by changing it.

6. C. *alatum* (W. & A.:) volubile glabriusculum, foliis (ramor. seniorum) basi cordato-auriculatis (ramul. juniorum florifer.) ovalibus cuspidatis basi cordatis vel emarginatis subtus glaucis, umbellis petiolum subæquantibus, pedicellis pedunculo longioribus, cor. st. corolla brevioris ore truncato crenato-10-lobato lobis alternis paullo minoribus, stigmate apiculato bifido, folliculis hinc planiusculis angulis alato-marginatis.—*Wight. ! cat. n.* 1552.—Asclepias truncata, *Roxb. ! in herb. Banks* (non *fl. Ind.*).——In Coromandelia.

The truncated crown and winged follicles will distinguish this from all the other species of the section.

7. C. *ovalifolium* (Wight:) volubile glabrum, foliis oblongo-ovalibus acuminatis, cymis multifloris, pedunculis petiolo longioribus, cor. st. corollam subæquante 10-fida, stigmate apiculato subemarginato.—Cynanchum, *Wall. ! Asclep. n.* 82.——In Penang; *Wallich.*

The oval not cordate, leaves and cymose inflorescence readily distinguish this species.—(*R. W.*)

8. C. *angustifolium* (W. & A.:) volubile glabriusculum, petiolis subhirsutis reflexis, foliis lineari-lanceolatis cuspidatis basi magis minusve cordatis (3–5 uncias longis 4–6 lineas latis), umbellis brevibus multifloris, cor. st. 10-lobæ lobis antheris oppositis corollam æquantibus late lineari-ovalibus alternis brevissimis inconspicuis omnibus truncatis emarginatisque, stigmate apiculato subemarginato.—*Wight. ! cat. n.* 1553.——In Coromandelia.

This was sent to Dr Wight by his collectors: the precise station is unknown. In several respects it agrees with *C. pauciflorum*, but is easily distinguished by the long (often 3 or 4 inches) and narrow linear-lanceolate slightly cordate leaves gradually tapering to a sharp point and there cuspidate, by the reflexed petioles, many-flowered umbels, and by the longer segments of the corona having the sides parallel in place of converging to a point: the form of the stigma is almost the same in both.

§ 4. *Gynostegium stipitatum. Cor. st. gamophylla tubulosa ad basin stipitis inserta, 5-fida, intus laciniis carinisve nullis.*

To this section, remarkable for the elongated torus or stalk to the gynostegium, belongs also a species from the Cape, which I had the opportunity of examining in Mr Lindley's herbarium.—(*R. W.*)

9. B. *gracillimum* (Wall.:) volubile glabrum, ramis gracilibus, foliis cordatis acuminatis basi incurvo-auriculatis, umbellis paucifloris, pedunculo petiolum subæquante pedicellis longiore, corollæ segmentis lanceolatis, gynostegii stipite cor. stamineam 5-fidam intus nudam superante, massis poll. apice affixis, (stigmate rostrato æquali 5-sulcato truncato antherarum membranas æquante.—ARN.)—*Wall. ! Asclep. n.* 86.——Sagaen; *Wallich.*—(*R. W.*)

I have added to the character proposed by Dr Wight, that of the stigma, which was omitted by him: this species differs therefore from the others of the genus, not only by the singularly stalked gynostegium protruded beyond the corona, but by the pretty long prismatical stigma: the membranous tip to the anthers is also long. The pollen masses are pendulous, ovate-oblong, and attenuated at the apex. I have not seen the species from the Cape noticed by Dr Wight but if it also possesses the rostrate, although included stigma, I conceive there is little doubt of the two forming a more distinct genus than any hitherto proposed to be separated from *Cynanchum* as limited by Mr Brown.—(*G. A. W.-A.*)

§ 5. *Cor. st. profunde 5-fida, laciniis antheris oppositis intus lacinula parallela auctis. Massæ poll. apicem vel infra apicem affixæ. Stigma subapiculatum.*

10. C. *auriculatum* (herb. Royle:) volubile, foliis cordato-ovatis acuminatis ciliatis basi auriculatis, umbellis cymæformibus longe pedunculatis paucifloris, corolla 5-partita, cor. st. profunde 5-fidæ laciniis obtusis gynostegium superantibus intus lacinula parallela auctis.—*Wall.! Asclep. n.* 137.——Kimour; *Royle.* Kamoun; *Wallich.*—(*R. W.*)

11. C. *Roylei* (Wight:) volubile ?, totum pubescens, foliis e basi late-cordata auriculata in apicem subulatum sensim attenuatis, umbellis longe pedunculatis multifloris, corolla 5-fida, cor. st. profunde 5-fidæ laciniis obtusis intus lacinula parallela auctus gynostegium æquantibus.—C. tomentosum, *herb. Royle.*——Kimour; *Royle.*

§ 6. *Cor. st. profunde 5-fida; laciniis simplicibus, massæ pollinis sub apice affixæ.—Plantæ decumbentes, vel erectæ, nunc subvolubiles, umbellis sessilibus vel subsessilibus.*

12. C. *glaucum* (Wall.:) herbaceum erectum glabrum vel pubescens, caulibus compressiusculis et subangulatis, foliis late ovatis vel oblongo-ovalibus vel lanceolatis obtusis subtus pallide glaucis, umbellis subsessilibus, corolla 5-fida segmentis intus hirsutis, cor. st. profunde 5-fidæ laciniis obtusis, massis poll. infra apicem affixis, stigmate obtuse apiculato.—*α, latifolium;* foliis late ovatis obtusis, umbellis multifloris.—C. glaucum, *Wall.! Asclep. n.* 133.—*β, oblongifolium;* foliis oblongo-ovalibus, umbellis paucifloris.—*Wall.! Asclep. n.* 132 (innomin.).—*γ, lanceolatam;* foliis oblongo-lanceolatis obtusis, umbellis sessilibus multifloris.—*Wight.! cat. n.* 1554 (ex herb. Madr.).—*Wall.! Asclep. n.* 134 (innomin.).——*α* et *β* ?, Nepal; *Wallich.*—*γ*, herb. Madras (absque local., vix Peninsula indigenum). Ad Simla; *Dom. Comit. de Dalhousie.* In itinere ad Cashmere; *Royle.*

13. C. *Kenouriense* (Wight:) herbaceum erectum, caule petiolisque breviter hirsutis, foliis ovalibus cuspidatis breviter petiolatis supra versus margines et in nervos venasque pubescentibus subtus nervis conspicuis venisque breviter hirsutis, umbellis paucifloris, cor. st. profunde 5-fidæ gynostegium æquantis laciniis obtusis sinu lato interjecto, massis poll. apici affixis, stigmate obtuso.——Kenour; *Royle.*

In this species the rhizoma is short, thick, and abrupt.—(*R. W.*)

14. C. *Arnottianum* (Wight:) herbaceum erectum glabrum, foliis oblongo-ovalibus obtusis emarginatisve mucronatis breviter petiolatis, umbellis sessilibus multifloris, corollæ 5-fidæ (extus fusco-purpureæ) segmentis intus albido-pilosis, cor. st. profunde 5-fidæ gynostegium æquantis laciniis obtusis sinu lato interjecto, massis poll. sub apice affixis, stigmate apiculato.——In itinere ad Cashmere; *Royle.*

The colour of the flowers will readily distinguish this from the other allied species.—(*R. W.*)

15. C. *nanum* (Ham.:) herbaceum erectum glabrum, foliis breviter petiolatis linearibus acutiusculis basi sub-auriculatis (4–5 uncias longis 3–5 lineas latis), umbellis subsessilibus multifloris involucratis, corollæ (majusculæ) 5-partitæ segmentis obtusis, cor. st. profunde 5-fidæ gynostegium æquantis laciniis obtusis, massis poll. apice affixis stigmate apiculato.—*Herb. Ham.!* n. 765; *Wall.! Asclep.* n. 141.——In arundinetis elevatis Camrupæ borealis; *Hamilton.*—(*R. W.*)

XXVII. DŒMIA. *R. Brown.*—Pergularia. *Lam. ill. t.* 176.

Corolla subrotata, tubo brevi. Corona staminea duplex; exterior brevis, 10-partita, laciniis alternis nanis; interior 5-phylla, foliolis basi solutis supra subulatis. Antheræ membrana terminatæ. Massæ pollinis compressæ, apice affixæ, pendulæ. Stigma muticum. Folliculi ramentacei. Semina comosa. —Caules volubiles. Folia opposita, cordata. Flores umbellati.

1. D. *extensa* (R. Brown:) fruticosa volubilis, foliis subrotundo-cordatis acuminatis acutis basi auriculatis pubescentibus subtus glaucis, pedunculis pedicellisque elongatis filiformibus, corolla margine ciliata.—*R. Br. in Wern. soc. mem.* 1. *p.* 50; *Roem. et Sch. syst. veg.* 6. *p.* 112; *Spr. syst.* 1. *p.* 853; *Wall.! Asclep. n.* 97; *Wight.! cat. n.* 1555.—Cynanchum extensum, *Ait. hort. Kew.* (*ed.* 1.) 1. *p.* 303; *Jacq. ic. var.* 1. *t.* 54.—C. cordifolium, *Retz. obs.* 2. *p.* 15. —Asclepias echinata, *Roxb. fl. Ind.* 2. *p.* 44; *in cat. merc. Ind. or. tab.* 616. ——In Peninsula Indiæ. Ad ripas Gangis, et ad Umbatta in Himalaiah; *Dom. Comit. de Dalhousie.*

In Dr Wallich's herbarium (Asclep. n. 98) there are specimens from Penang of perhaps another species, but they are too imperfect to be characterised or even with certainty proved distinct.

XXVIII. SARCOSTEMMA. *R. Brown.*

Corolla rotata. Corona staminea duplex; exterior cyathiformis vel annularis, crenata; interior 5-phylla, exteriorem superans, foliolis carnosis. Antheræ membrana terminatæ. Massæ pollinis apice affixæ, pendulæ. Stigma apiculatum vel muticum. Folliculi graciles, læves. Semina comosa.—Frutices volubiles vel decumbentes. Caules aphylli articulati, vel foliis oppositis distantibus. Umbellæ laterales vel terminales.

1. S. *viminale* (R. Brown:) volubile aphyllum, umbellis terminalibus, calyce pedicellisque canescentibus, cor. st. exterioris 10-plicatæ 10-dentatæ dentibus subæqualibus, interioris foliolis planiusculis antheras fere superantibus, stigmate apiculato bifido.—*Wall.! Asclep. n.* 99. *c*; *Wight.! cat. n.* 1556. ——In Indiæ Peninsula.

This and the following species appear to be so confused by authors under the name of *Cynanchum viminale*, that we find it impossible to refer to their works. Mr Brown's character of his *S. australe* seems likewise applicable to them all, nor do we know how it differs: perhaps even it contains two species, one with terminal, the other with lateral flowers.

2. S. *Brunoniana* (W. & A.:) volubile aphyllum, umbellis lateralibus sessilibus, calyce pedicellisque canescentibus, cor. st. exteriore subplicata ore 10-crenata crenis alternantibus subobsoletis, interiore antheris breviore, stigmate apiculato subintegro.—*Wight. cat. n.* 1557.——In Coromandelia australi.

3. S. *brevistigma* (W. & A.:) volubile aphyllum, umbellis terminalibus vel ramulos breves laterales terminantibus, calyce pedicellisque glabris, cor. st. exteriore 10-plicata 10-crenata, interioris foliolis dorso gibbosis gynostegium æquantibus, stigmate mutico.—*Wight.! cat. n.* 1558.—S. viminale, *Wall.! As-*

clep. n. 99. *t.*—Asclepias acida, *Roxb. fl. Ind.* 2. *p.* 31.—A. aphylla, *Roxb. in cœt. merc. Ind. or. tab.* 607.——In Coromandelia.

We have referred here to Roxburgh, as his drawing resembles this in the stigma, and rather more in habit than the two preceding: but the corona is not accurately represented.

XXIX. PENTASACHME. *Wall.*

Corolla subrotata, profunde 5-fida; fauce squamulis 5 segmentis alternis coronata: æstivatio valvata. Corona staminea nulla. Antheræ liberæ, apice simplices vel acuminatæ. Massæ pollinis pendulæ, supra medium affixæ, apice coarctato pellucido. Stigma muticum vel apiculatum.—Herbæ graciles erectæ subramosæ. Folia opposita, plana, membranacea. Umbellæ subsessiles, pauciflorae. Corollæ segmenta tubo duplo triplove longiora, anguste linearia.

A genus allied in many points to *Leptadenia*, but distinguished by the form of the corolla, the pendulous pollen-masses, and by the very different habit. The fruit is unknown.

1. P. *caudatum* (Wall.:) foliis lanceolatis longe attenuatis, corollæ segmentis subulatis faucis squamulis majusculis 4-fidis, antheris obtusis, stigmate mutico.—*Wall.! Asclep. n.* 75.——Silhet, in montibus Lahore; *Wallich.*—(*R. W.*)

2. P. *Wallichii* (Wight:) foliis lanceolatis acuminatis, umbellis subsessilibus paucifloris, corollæ segmentis linearibus obtusis faucis squamulis parvis glandulæformibus, antheris acuminatis, stigmate subrostrato.—*Wall.! Asclep. n.* 74 (innomin.)——Silhet; *Wallich.*—(*R. W.*)

II. SECAMONEÆ.

Massæ pollinis 20, læves, quaternatim (duabus antheris pertinentes) affixæ apici corpusculorum exsulcorum stigmatis. Filamenta connata, extus appendiculata.—*R. Brown.*

XXX. SECAMONE. *R. Brown.*

Corolla rotata. Corona staminea 5-phylla; foliolis lateraliter compressis margine longitudinali affixis, aversis, simplicibus. Massæ pollinis 20, erectæ, quaternatim affixæ apici corpusculi singuli exsulci stigmatis. Stigma apice coarctato. Folliculi læves.—Frutices erecti vel volubiles, glabri. Folia opposita. Cymæ dichotomæ, interpetiolares. Flores minuti.

1. S. *emetica* (R. Brown:) volubilis glabra, foliis ex ellipticis in anguste lanceolata, cymis folio brevioribus pauci vel multifloris corolla glabra, cor. st. foliolis cultriformibus gynostegio subdimidio brevioribus, folliculis gracilibus apice attenuatis.—*Br. in Wern. soc. mem.* 1. *p.* 55; *Roem. et Sch. syst. veg.* 6. *p.* 124; *Spr. syst.* 1. *p.* 837; *Wall.! Asclep. n.* 102; *Wight.! cat. n.* 1559.—Periploca emetica, *Retz. obs.* 2. *p.* 14; *Willd. Phyt.* 1. *p.* 6, *t.* 5. *f.* 3; *sp.* 1. *p.* 1250.—P. cymosa, *Roxb. mst.*——In Peninsulæ provinciis australioribus, ad radices montium in dumetis, frequens.

Roxburgh considers the *P. emetica* of Retz and Willdenow to be the narrow-leaved state of his *Ascl. pseudosarsa*, but to this we cannot consent, because both his broad and narrow leaved forms of that plant belong to *Hemidesmus Indica:* nor does it appear that Roxburgh ever found the present species, at least it does not correspond with any of his descriptions, and seems confined to a region further south than any which he had an opportunity of examining. His manuscript name, above referred to, he appears only to have attached to dried specimens sent him by the missionaries after he went to Bengal. Mr Brown (Wern. soc. mem. 1. p. 56) mentions another species discovered by Roxburgh, but this we suspect to be our *Toxocarpus Roxburghii.*

2. S. *Finlaysoni* (Wight:) volubilis glabra, foliis ovalibus acuminatis subsessilibus, cymorum folio longiorum ramis flexuosis, corolla glabra, cor. st. foliolis gynostegium subæquantibus.—*Wall.! Asclep. n.* 103 (innomin.)—— .... ; *Finlayson.*

XXXI. TOXOCARPUS. *W. & A.*

Corolla rotata, 5-fida. Corona staminea 5-phylla; foliolis dorso planiusculis, intus lacinula auctis. Massæ pollinis 20, erectæ, quaternatim affixæ apici corpusculi singuli exsulci. Stigma rostratum, rarius apiculatum, indivisum. Folliculi læves, divaricati. Semina comosa.—Frutices volubiles. Corymbi interpetiolares, sæpius oppositi, dichotomi, patentes, multiflori.

1. T. *Kleinii* (W. & A.:) caulibus glabris, ramulis junioribus pubescentibus, foliis ellipticis subiter ac breve acuminatis, corymbis subsessilibus ramis divaricatis folio longioribus, floribus pedicellatis, corollæ segmentis ligulatis glabris fauce subpilosa, cor. st. foliolis ovatis bidentato-truncatis apice lacinulam interiorem planam linearem longe exsertam antheras superantem apice subtridentatam gerentibus, stigmate rostrato apice breviter bifido corollæ tubum paullo superante, folliculis arcuato-reflexis.—*Wight.! cat. n.* 1560.—Periploca glabra, *Roxb.! in herb. Banks.*—Echites racemosa, *herb. Madr.!*—*Wall.! Asclep. n.* 113. *a* (innom.)——In sepibus apud Vellangany prope Nepatam, frequens; *Wight.*

To this is closely allied *T. Wightianus*, Hook. and Arn., from China.

2. T. *Roxburghii* (W. & A.:) ramulis ferrugineo-pubescentibus, foliis late ovalibus subiter ac breve acuminatis, corymbis breviter pedunculatis ramis divaricatis folio subæqualibus, floribus subsessilibus, corollæ fauce hirsuta segmentis ligulatis glabris, cor. st. foliolis ovatis acutiusculis lacinula interiore brevi crassa acuta vix exserta antheras æquante, stigmate rostrato corollæ tubum æquante contorto.—*Wight.! cat. n.* 1561.—Asclepias longistigma, *Roxb. fl. Ind.* 2. *p.* 46; *in cœt. merc. Ind. or. mus. tab.* 619.—A. micrantha, *Roxb.! in herb. Banks* (non *fl. Ind.*)—Echites dichotoma, *herb. Klein! et Rottl.!; Wall.! Asclep. n.* 113. *b.*——Circars; *Roxburgh.*

The flowers are much smaller than in the last species.

3. T. *crassifolius* (Wight:) ramulis parce pubescentibus, foliis breve petiolatis ovalibus acuminatis coriaceis glabris, corymbis paniculiformibus sessilibus ramis elongatis parce ramosis calyceque fusco-pubescentibus, floribus sessilibus fasciculatis parvis, corollæ segmentis reflexis dense albido-villosis, cor. st. foliolis lateribus reflexis superne cohærentibus carinamque dorsalem efformantibus lacinula interiore acuminata longe exserta stigma subæquante, stigmate rostrato gynostegioque corollæ tubum superantibus.—Secamone crassifolia, *Wall.! Asclep. n.* 101.——In Silhet; *Wallich.*—(*R. W.*)

4. T. *laurifolius* (Wight:) glaber, foliis ovalibus nunc obtusis sæpius acuminatis coriaceis, corymbis sessilibus divaricatim trichotomo-ramosissimis, floribus valde numerosis parvis, corollæ segmentis reflexis intus albido-barbatis, cor. st. foliolis subcarnosis brevibus rotundato-ovatis acutiusculis lacinula interiore breviter exserta gynostegium subæquante, antherarum apicibus majusculis latis subcristatis supra stigma obconico-apiculatum obtusum inclusum incurvatis, folliculis gracilibus horizontaliter divergentibus.—Asclepias laurifolia, *Roxb. fl. Ind.* 2. *p.* 49; *in cœt. merc. Ind. or. mus. tab.* 1814; *Wall.!*

Asclep. n. 100.—A. micrantha, *Roxb.! in herb. Banks!; fl. Ind.* 2. *p.* 50 (planta junior).——In Chittagong et Tipperah, atque in montosis e Benghala orientem versus tendentibus; *Roxburgh.* In Hindoostania apud Cawnpore; *Hardwick.*

XXXII. GONIOSTEMMA. *Wight.*

Corolla rotata, 5-partita. Corona staminea gamophylla tubulosa 5-angulata 5-loba, gynostegii basi adhærens. Massæ pollinis 20, erectæ quaternatim affixæ apici corpusculi singuli exsulci. Stigma rostratum. Folliculi .... Frutex volubilis, cortice verruculoso. Folia opposita, oblongo-elliptica, basi apice acuminata, glabra, supra nitida. Cymæ interpetiolares, paniculæformes, laxæ, multifloræ. Corollæ segmenta ligulata, intus ad medium pubescentia, superne glabra. Cor. st. gynostegio brevior, carnosa, lobis intus nudis. Stigma cylindraceum, corolla subdimidio brevius.

This genus is allied to both *Secamone* and *Toxocarpus*, but differs from both in the structure of the corona and also in habit. Brown's character of *Secamone*, however, was perhaps intended to include all the three.

1. G. *acuminatum* (Wight.)—*Wall.! Asclep. n.* 127 (innomin.)——Silhet; *Wallich*, 1828.

III. PERIPLOCEÆ. *R. Brown.*

Massæ pollinis 5–20, granulosæ, solitariæ vel quaternatim appendiculo dilitato solitario (e binis cohærentibus composito) ex apice corpusculi singuli exsulci orto stigmatis demum applicitæ. Filamenta partim vel omnino distinctæ. Antheræ conniventes, tholi modo stigma contegentes.

In all the genera of this group which we have examined (except perhaps in *Cryptostegia*, in which our specimens were not sufficiently advanced), the corpuscle is dilated at the base into a kind of gland similar to those found at the base of the pollen-masses of an *Orchis;* and in all, with the above exception, both the corpuscle and its appendage are at length readily detached from the stigma.

XXXIII. HEMIDESMUS. *R. Brown.*

Corolla rotata, squamis 5 muticis sub sinubus insertis. Filamenta basi connata, superna distincta, tubo corollæ inserta. Antheræ cohærentes, a stigmate liberæ, imberbes, apice simplices. Massæ pollinis 20, granulosæ, quaternatim appendiculo reniformi solitario corpusculi singuli applicitæ. Stigma planiusculum, muticum. Folliculi cylindracei, divaricatissimi, læves. Semina comosa.—Frutices volubiles. Folia opposita, supra nitida. Cymæ interpetiolares. Flores parvuli. Corolla carnoso-coriacea.

The anthers are 4-celled: cells soon bursting and becoming expanded, so that the anthers appear as if furnished with a broad membranous appendage or wing on each side, forming a dome over the stigma, covering and apparently defending from wet the pollen-masses. Pollen-masses granular, four to each anther: these remain for some time in their cells, but at length adhere to the viscid dilated corpuscular appendage. The spherules of which the masses are composed do not appear as if united by fours into distinct granules, as in *Periploca*, but seem rather to cohere together indeterminately: thus, if the mass be bruised on the field of the microscope, pieces may be observed of a roundish shape, others linear; some composed of 2 or 3 spherules, others of 4, 5, or more, either heaped together irregularly or chain-like. Ovaries 2. Style very short. Stigma large, peltate, orbicular, slightly convex, glabrous. Processes 5, alternate with the anthers, at first mere glandular secretions across the edge of the stigma, afterwards of a firm texture, and readily separating from it; in this last stage the corpuscle, or lower part, is linear-oblong, dilated at the base, into a discoid gland, which is broader and paler than the corpuscle; it is not marked by a furrow. The upper part, which obviously corresponds to the appendiculus seen in most other Asclepiadeæ, is attached to the corpuscle by a mere point, apparently by a kind of joint, and is broadly reniform and slightly emarginate; this is marked by a longitudinal line, indicating that it is formed by the union of two pieces. After this part has received both pollen-masses from the adjacent cells of the two contiguous anthers, which it does not seem to do till after the flowers are expanded, the lobes at its base bend upwards, and render it cucullate, so as to resemble a little cup containing 4 pollen masses, while the corpuscle serves as a stalk to it: it seems afterwards to become again flattened out, when the pollen-masses disappear.

1. H. *Indicus* (R. Brown:) glaber, foliis ex cordato-ovatis cuspidatis in anguste linearia acuta sæpius oblongo-lanceolatis, cymis sæpius subsessilibus nunc pedunculatis, corollæ squamis obtusis tubo e basi usque ad apicem cohærentibus, folliculis gracilibus rectis.—*Hort. Kew.* 2. *p.* 75; *Roem. et Sch. syst. veg.* 6. *p.* 126; *Spr. syst.* 1. *p.* 836; *Ham.! herb. n.* 761, 764; *Wall.! Asclep. n.* 104, 105; *Wight.! cat. n.* 1562.—Periploca Indica, *Willd. sp.* 1. *p.* 1251; *Roxb. in cæt. merc. Ind. or. mus. tab.* 212.—Asclepias pseudosarsa, *Roxb. fl. Ind.* 2. *p.* 39 (excl. syn. plurimis). *Rheed. Mal.* 10. *t.* 34; *Burm. Zeyl. p.* 187. *t.* 83. *f.* 1; *Pluk. t.* 359. *f.* 2 (foliis latis), *et* 361. *f.* 1 (fol. angust.).——In Indiæ Peninsula, ubique. In dumetis saxosis Magadhæ; *Hamilton.*

A very polymorphous species, perhaps to be afterwards divided when we become better acquainted with the fruit of the different varieties; but at present we can find no fixed character by which to separate the broad from the narrow leaved ones, the foliage presenting every intermediate form: we have, however, seen no leaves so long and narrow as in Rheede's figure, although we have little doubt, from his description, that it really belongs to this species; in that respect it approaches *H. Wallichii.*

2. H. *pubescens* (W. & A..) ramulis gracilibus hirsuto-pubescentibus, foliis lanceolatis acutis pedunculis calyceque pubescentibus, cymis sessilibus, corollæ squamis tubo affixis.—*Wight.! cat. n.* 1563.—*Wall.! Asclep. n.* 105. *d* (innom.).——Apud Vendalore; *Wight.*

3. H. *Wallichii* (Wight:) glaber, foliis anguste lineari-lanceolatis acutis, cymis paucifloris, corollæ squamis subulatis a tubo liberis basi filamentorum cohærentibus.—*Wall.! Asclep. n.* 106 (innom.).——In collibus Prome; *Wallich.*

Besides the above, Mr Brown mentions two other unpublished Indian species "very nearly related" to *H. Indicus*; with these I am unacquainted, unless they belong to some of the forms of that plant.—(*R. W.*)

XXXIV. BRACHYLEPIS. *W. & A.*

Corolla subrotata, 5-partita, segmentis patulis; fauce squamulis 5 brevibus truncatis planis muticis segmentis alternantibus coronata; tubo brevi intus annulo lato carnoso adnato penitus vestito. Filamenta brevissima, lata, distincta, fauci summo cum squamulis inserta. Antheræ basi margini stigmatis adglutinatæ, apice simplices cohærentes cæteroquin distinctæ, imberbes. Massæ pollinis 20, granulosæ, appendiculo corpusculi singuli dilatato soluto applicitæ. Stigma muticum. Folliculi divaricatissimi cylindracei, læves.—Frutex volubilis. Ramuli pubescentes. Folia opposita, ovalia, abrupte acuminata, juniora subpubescentia, adulta glabra supra nitida subtus parallele nervosa. Cymæ interpetiolares, parvæ, tomentosæ, foliis breviores. Flores parvi, congesti. Corolla extus hirsuta, intus glabra, subrotata segmentis tantum patulis vel potius subcampanulata.

1. B. ***nervosa*** (W. & A.)—***Wight.! cat. n.*** 1565.—Streptocaulon nervosum, *Wight. mst.*—*Wall.! Asclep. n.* 107 (innom.).——Ad Neelgherry; *Noton; Wight.*

XXXV. DECALEPIS. *W. & A.*

Corolla subrotata, 5-partita; fauce squamis 5 oblongo-ovalibus planis muticis segmentis alternantibus coronata; tubo intus ad basin squamis 5 linearibus obtusis segmentis oppositis instructo. Filamenta distincta, fauci summo cum squamis 5 superioribus inserta iisdemque ad medium cohærentia. Antheræ basi margini stigmatis adglutinatæ, superne distinctæ, appendiculo subdilatato terminatæ, imberbes. Massæ pollinis granulosæ, solitariæ (e quatuor confluentibus), appendiculo dilatato corpusculi singuli soluto applicitæ. Stigma muticum.—Frutex volubilis?, glabriusculus. Folia opposita, ovalia, obtuse acuminata. Cymæ parvæ, longiuscule pedunculatæ et foliis breviores, ramis racemiformibus. Flores parvi. Corolla limbus intus villosus.

After Dr Wight left with me the manuscript description of the only species of this genus, I perceived that, in the catalogue sent along with the *Asclepiadeæ* from the Linnean Society, it was marked as being obtained from the Madras, and not from Hamilton's, herbarium, as Dr W. had supposed, and which circumstance had obviously suggested to him the specific name. It being thus presumed to be a Peninsular plant, I was induced to examine the flowers myself, and have in consequence not only made some slight alterations on the character, but separated it from *Streptocaulon*, with which my friend seemed inclined for the present to combine it.—(*G. A. W.-A.*)

1. D. *Hamiltonii* (W. & A.)—*Wight.! cat. n.* 1566.—Streptocaulon Hamiltonii, *Wight. mst.*—Apocynum reticulatum, *herb. Madr.; Wall.! Asclep. n.* 139.

The locality is unknown.

XXXVI. STREPTOCAULON. *W. & A.*

Corolla rotata, 5 partita; fauce squamis 5 brevibus segmentis alternantibus aristatis coronata; aristis flexuoso-erectis, filiformibus, apice rectis. Filamenta distincta, tubo inserta. Antheræ basi margini stigmatis adhærentes, superne liberæ, apice simplices, imberbes. Massæ pollinis granulosæ, solitariæ (e quatuor conflatæ), appendiculo dilatato soluto corpusculi singuli applicitæ. Stigma muticum. Folliculi cylindracei, divaricatissimi, læves. Semina comosa.—Plantæ volubiles rarissimeve decumbentes, sæpius pubescentes vel tomentosæ rarius glabræ. Folia opposita. Cymæ interpetiolares, sæpius patentes et ramosissimæ. Flores parvi.

This genus is quite intermediate between *Periploca*, as limited by Mr Brown, and his *Gymnema*. We would have preferred uniting it to the former, from which it seems only to differ by the beardless anthers, had not Mr Brown and all succeeding botanists laid considerable stress on that character: the species of *Periploca*, however, are natives of the south of Europe, Syria, north of Africa, and Sierra-Leone: those of this new genus, of East India alone. From *Gymnanthera* they differ by several characters, by the stigma, pollen-masses, and shape of the corolla and of its scales. Whether all or any of Poiret's Indian species, as might be inferred from his generic character, belong to this genus, we have no means of determining; nor do we know what his *P. Mauritiana* can be, the synonyms adduced being very different from each other; thus, Pluk. t. 336. f. 7. is *Tylophora asthmatica*, and Rheed. Mal. 9. t. 11. is *Cryptolepis Buchanani*, Roem. and Sch., which belongs to the *Apocyneæ*.

1. S. *tomentosum* (Wight:) ramis pubescentibus, foliis obovato-lanceolatis acuminatis basi subcordatis supra hispidis subtus tomentosis, cymis magnis diffusis, floribus breviter pedicellatis ad apices ramorum congestis, calyce pedicellisque dense pubescentibus, staminum filamentis brevissimis.—Perioplo-

ca tomentosa, *Wall. Asclep. n.* 125, 126.——Taong-Dong; *Wallich.* Cochin-china; *Finlayson.*—(*R. W.*)

2. S. *Kleinii* (W. & A.:) ramis pubescentibus, foliis subsessilibus cuneato-oblongis obovatisve mucronatis basi cordatis supra hispidulis subtus molliter albido-tomentosis, cymis diffusis folio brevioribus, calyce pedicellisque densc pubescentibus.—*Wight.! cat. n.* 1564.—*Herb. Madr.; Wall. Asclep. n.* 157 (absque nom. et hab.).

This we only know by means of the specimens from the Madras herbarium: unfortunately the locality is not mentioned.

3. S. *Wallichii* (Wight:) ramis pubescentibus, foliis ellipticis basi apice attenuatis cuspidatis supra molliter pubescentibus, subtus fulvo-velutinis, cymis valde diffusis ramosisque folio longioribus multifloris ramulis ultimis racemiformibus, pedicellis filiformibus calyceque glabris.—*Wall. Asclep. n.* 124 (innom.).——Penang et Singapore; *Wallich; Finlayson.*—(*R. W.*)

4. S. *extensum* (Wight:) glabrum, foliis ovalibus acuminatis, cymis paniculæformibus folio multo longioribus, floribus sparsis pedicellatis, pedicellis filiformibus glabris.—Periploca extensa, *Wall. Asclep. n.* 128.——Rangoon et Tavoy; *Wallich.*—(*R. W.*)

5. S. *sylvestre* (Wight:) procumbens vel subvolubilis, caulibus subvillosis, foliis subsessilibus rotundato-ovatis cuspidatis basi cordatis utrinque subtus præcipue brevissime subvelutinis, cymis folio brevioribus paucifloris, pedicellis calyceque pubescentibus.—Periploca? sylvestris, *Ham. herb. n.* 763; *Wall. Asclep. n.* 140.——Sanashygota; *Hamilton.*—(*R. W.*)

6. S. *calophyllum* (Wight:) glabrum, foliis anguste lanceolatis longe attenuatis transversim venosis margine subundulato, cymis subsessilibus paucifloris, floribus breviter pedicellatis, folliculis gracilibus attenuatis (circiter 6 uncias longis).—Periploca? *Wall. Asclep. n.* 123.——Nepal; *Wallich.*

This plant presents a very different aspect from any of the other species, but the structure of the flower appears quite the same.—(*R. W.*)

XXXVII. FINLAYSONIA. *Wall.*

Corolla rotata; fauce tuberculis 5 aristatis segmentis alternantibus coronata, aristis capillaceis erectis apice uncinatis. Filamenta distincta, fauci intra tubercula inserta. Antheræ basi medio stigmatis adglutinatæ, apiculo late-ovato terminatæ, imberbes. Massæ pollinis 20, granulosæ, appendiculo dilatato corpusculi singuli gracilis quaternatim demum applicitæ. Stigma magnum, ovatum, basi 5-gonum. Folliculi divaricatissimi, ventricoso-ovati, apice rostrato-uncinati, læves. Semina parce carnosa.—Frutex volubilis, carnosus, lactescens, glaber. Folia opposita, petiolata, obovata, obtusissima, nunc retusa cuspidula latiuscula acuta. Corymbi majusculi, multiflori, folio breviores, fructiferi valde elongati. Flores parvi. Corollæ extus glabræ segmenta rotundato-ovata intus hinc nuda, inde albido-villosa.

It rarely happens that all the pollen-masses are found on the corpuscle or its appendage, the upper ones of each cell of the anther being often found remaining, at least for a very long while, in their original situation.

1. F. *obovata* (Wall.)—*Wall.! pl. As. rar.* 2. *p.* 48. *t.* 162.—Gurua obovata, *Ham. herb. n.* 757.——Ad æstuaria Gangetica; *Wallich; Hamilton.*

XXXVII. CRYPTOSTEGIA. *R. Brown.*

Calyx 5-partitus; laciniis elongatis, lanceolatis, margine undulatis. Corolla campanulato-infundibuliformis; tubo intus squamis 5 inclusis angustis bipartitis (segmentis subulatis) antheras obtegentibus iisdemque oppositis instructo. Stamina inclusa. Filamenta distincta, brevissima, imo tubo inserta. Antheræ inclusæ, basi margini stigmatis adhærentes. Massæ pollinis solitariæ (é granulis liberis permultis coacervatis), appendiculo spathulato tandem soluto corpusculi singuli utrinque margini imo antherarum proximarum adglutinati demum applicitæ. Stigma globoso-conicum, apice obscure biapiculato. Folliculi divaricatissimi, acute-triquetri, apice obtuse attenuato incurvo. Semina comosa.—Frutex volubilis, glaber, lactescens. Folia opposita, breviter petiolata, elliptica, obtuse acuminata, utrinque nitida, subtus minute fusco-reticulata. Corymbi trichotomi, terminales. Flores magni, speciosi, rubelli. Corolla æstivatione contorta.

1. C. *grandiflora* (R. Brown).—*Br. in bot. reg. t.* 435; *Wall.! cat. n.* 1635; *Wight.! cat. n.* 1660.—Nerium grandiflorum, *Roxb. fl. Ind.* 2. *p.* 10; *in cæt. merc. Ind. or. mus. tab.* 1957.

Roxburgh states that the native country of this plant is the Peninsula of India, but he appears to have presumed so only from its having been sent him from thence by Heyne in 1804. Dr Wight has never seen it except in a state of cultivation.

NOTE.

We have in the course of these descriptions referred to Vahl's Memoir in the 6th volume of the Transactions of the Natural History Society of Copenhagen (Skrivter af Naturhistorie-Selskabet, Kiobenh.), but as that volume is so extremely scarce as to be in very few botanists' possession, we shall here extract from it the descriptions of the genera and species belonging to *Asclepiadeæ*, premising, that the page following the specific appellation is that of the volume alluded to, and the name following the genus is that to which we are disposed to refer it.

COLLYRIS. (Dischidia, *Br.*)

Contorta. Cor. globosa: fulcrum erectum bifidum; laciniis reflexis, clavatis, ad singulum angulum corporis truncati.

1. C. *major* (p. 110:) foliis orbiculatis.—Pustula arborum, *Rumph. Amb.* 5. *p.* 473. *t.* 175. *f.* 3.

HAB. In sylvis aridis et collinis prope Malaccam et Solangohr. *Koenig.*

Caules scandentes, truncis arborum affixi mediantibus squamis oppositis, reniformibus, albidis, latere interiore radicantibus: præterea *Vesiculæ* plures, sessiles, ovato-oblongæ, bipollicares vel ultra, obtusissimæ, extus glabræ, cicatricibus magnis notatæ, intus parum inflatæ, purpureæ, fibra longitudinali ramosissima basi vesiculæ inserta, sæpe aqua dulci plenæ. *Rami* teretes, flexuosi, radicantes, inferne e tuberculis pluribus nodosi, superne æquales. *Folia* opposita, remotissima, petiolata, vix pollicaria, subrotunda, acuta, interdum parum oblonga, carnosa. *Flores* axillares. *Calyx* parvus, quinquefidus. *Corolla* globosa, ore quinquefido: laciniæ erectiusculæ. *Corpus* truncatum apice convexum, quinquangulare, magis lineare quam in affinibus, angulis fissis, marginibus parum callosis: ad singulum angulum *Fulcrum* erectum, apice bifidum; laciniis recurvatis, clavatis, longitudine corpusculi truncati. *Hypomochlia* parva, uti in affinibus inter squamulas. *Antheræ* oblongo-clavatæ, planæ, callosæ, flavæ. *Stigmata* duo in centro corporis truncati, eoque arcte connata et parum longiora. *Folliculi* lineari-oblongi, rostrati glabri, læves, latere superiore planiusculi, apicibus incurvatis. *Semina* comosa.

Nummulariæ, Rumph. 5. tab. 175, genere differunt, cum tam ad inflorescentiam quam quoad figuram corollarum potius insequentis generis species sint, quamvis eodem nomine insignentur a Rumphio.

2. C. *minor* (p. 111); foliis oblongo-ovatis.—Nummularia lactea minor, *Rumph. Amb.* 5. *p.* 472. *t.* 176. *f.* 1.—An acinos multis effigie ex cauliculorum geniculis radicosa, herba capillaris ex insula Cheusan, *Pluk. amalth.* 4. *t.* 354. *f.* 2.

HAB. In India orientali; *Koenig.*

Rami filiformes, teretes, glauci, ad folia radicantes, flexuosi. *Folia* sessilia, opposita, vix unguicularia, quadruplo minora quam in præcedente, integerrima, utrinque acuta, carnosa, enervia, avenia, subtus glauca. *Flores* axillares.

SPERLINGIA. (Hoya, *Br.*)

Contorta. Nectarium stellula simplex genitalia tegens. Petala 5, apice marginibusque inflexis triangularia, plana, conduplicata.

1. S. *verticillata* (p. 113:) foliis quaterno-verticillatis.

HAB. In India orientali.

Caulis scandens, radicans, lignosus, teres, cortice cinereo, glaber, lævis. *Folia* petiolata, oblonga, acuminata, acuta, integerrima, carnosa, crassa, glabra, avenia, enervia, glabra. *Petioli* fere pollicares, succulenti, plano-depressi, lati, basi articulati. *Pedunculi* laterales, teretes, glabri. *Flores* umbellati, copiosi: *pedicelli* pollicares, filiformes, compressi, uniflori. *Receptaculum* commune, primum hemisphæricum, demum cylindricum, pedunculo parum crassius, e casu pedicellorum conferte cicatricatum. *Calyx* pentaphyllus; foliola oblonga. *Petala* quinque, apice margine inflexo conduplicata, triangularia, margine exteriore retuso-subbicornia; explanata oblonga, utrinque acuta. *Corpus* truncatum stellula quinque radiata; radii lanceolati, longitudine petalorum complicatorum, cum petalis alternantes, media linea elevata, ad cujus latera foveola longitudinalis, flavescentes, nitidi. *Stamina* non vidi. *Germina* duo sub centro corpusculi. *Folliculus* sex-octopollicaris, glaber, lævis. *Semina* comosa.

2. S. *opposita* (p. 114); foliis oppositis.—Nummularia lactea major, *Rumph. Amb.* 5. *p.* 470. *t.* 175. *f.* 1.

HAB. In India orientali: ex herbario bibliothecæ Christianiensis.

Similis in omnibus præcedenti, diversa: foliis oppositis; petiolis angustioribus, subtus convexis, supra canaliculatis.

No. III.

CYPERACEÆ INDICÆ,

PRÆCIPUE JUXTA HERBARIA WIGHTII, WALLICHII, ROYLEI ET LINDLEYI, AB PROF. CH. G. NEES AB ESENBECK DESCRIPTÆ.

ORDO GENERUM.

Trib. I. CYPEREÆ.

Hermaphroditæ. Spiculæ distiche imbricatæ, uni-multifloræ, plerumque nudæ, rarius in spicam undique imbricatam collectæ sub singula squama singulæ. Perigynium nullum, aut (in speciebus exoticis) setaceum, aut cyathiforme, squamulæ interiores rhachillæ adnatæ aut nullæ. Caryopsis stylo discreto non rostrata, mutica, in paucis acumine rostriformi prædita.

A. *Spiculæ liberæ, plurifloræ: perigynium nullum.*

1. Cyperus (*Linn.*)—Rhachilla squamulis interioribus adnatis persistentibusque marginata. Stylus bi-trifidus. Squamæ spiculæ inferiores cum reliquis deciduæ. Stamina decidua.

2. Papyrus (*Link.*)—Squamulæ interiores apice aut totæ liberæ. Stylus trifidus.

3. Mariscus (*Vahl.*)—Spiculæ paucifloræ. Rhachilla squamulis interioribus adnatis persistentibusque alata. Stylus trifidus. Spicula a squamis duabus infimis decidua.

B. *Spiculæ liberæ, subunifloræ.*

4. Kyllingia (*Linn.*)—Stylus elongatus bifidus. Caryopsis compressa.

5. Courtoisia (*N. ab E.*)—Stylus ad basin fere tripartitus. Caryopsis trigona, libera.

6. Remirea (*Aubl.*)—Stylus trifidus. Caryopsis trigona, squamula interiore tunicata.

Trib. II. HYPOLYTREÆ.

Hermaphroditæ. Squamæ spiculæ compositæ undique imbricatæ. Spicula partialis aut disticha pluriflora, aut uniflora; squamulis tum flosculum unum in plurimis hermaphroditum, in nonnullis declinem valvatim aut circulo ambientibus, haud setiformibus aut cyathoideis.

A. *Spicula partialis e squamis pluribus constructa, pluriflora.*

7. ANOSPORUM (*N. ab E.*)—Spiculæ in spicas compositas undique imbricatas collectæ. Perigynium utriculare, stipitem caryopseos construens, superne evanescens. Stylus simplex.

B. *Spicula uniflora, e squamis paucis quibusdam ternatim alternis constructa.*

8. HEMICARPHA (*N. ab E.*)—Perianthium proprium univalve, squamæ communi oppositum.

9. LIPOCARPHA (*R. Br.*)—Perianthium proprium bivalve, squamæ communi subparallelum.

10. HYPOLYTRUM (*Rich.*)—Perianthium proprium bivalve (rarius distichoquadrivalve valvulis inferioribus sterilibus), squamæ communi contrarium. Stamina 2–3.

11. FUIRENA (*Rottl.*)—Perianthium trisepalum aut hexasepalum, et tum vero sepalis alternis setiformibus. Stamina tria.

TRIB. III. SCIRPEÆ.

Hermaphroditæ. Spiculæ undique imbricatæ. Squamulæ perigynii aut nullæ, aut angustæ, setiformes aut filiformes, flosculum cingentes, aut breve cyathiforme membranaceum.

A. *Squamæ bi-trifariæ.*

a. *Perigynium nullum.*

12. DICHOSTYLIS (*P. de B.* ex parte).—Spiculæ subtrifariam imbricatæ, molles, pauciflores. Stylus bifidus, basi subbulbosa persistente.

13. ABILDGAARDIA (*Vahl.*)—Spiculæ torsione rhachillæ trifariæ. Stylus trifidus, basi bulbosus, deciduus.

b. *Perigynium.*

14. CHÆTOCYPERUS (*N. ab E.*)—Spicula solitaria, disticha. Perigynii setæ sex, cartilagineæ, retrorsum hispidæ. Stylus trifidus, basi bulbosa ovario articulo inserta persistente.

B. *Squamæ plurifariæ.*

a. *Perigynium nullum.*

15. ECHINOLYTRUM (*Desv.*)—Stylus bifidus, basi bulbosus deciduus. Caryopsis grosse echinata.

16. FIMBRISTYLIS (*Vahl.*)—Stylus bifidus, compressus, basi bulbosus, deciduus. Caryopsis compressa, non echinata.

17. TRICHELOSTYLIS (*Lestib.*)—Stylus trifidus, basi bulbosus, deciduus. Caryopsis trigona. (Squamæ 4–8-fariæ, carinatæ).

18. ISOLEPIS (*R. Br.*)—Stylus trifidus, basi subæqualis, deciduus. Caryopsis trigona aut plano-conexa.

19. ELEOGITON (*Link.*)—Stylus bifidus, basi cum fructu continuus, post casum mucronulum relinquens. Caryopsis compressa, pedicellata. Spicula solitaria, terminalis.

b. *Perigynium filiforme, setaceum, capillare.*

20. MALACOCHÆTE (*N. ab E.*—Pterolepis, *Schrad.*)—Perigynii setæ lineares, membranaceæ, pubescentes aut ciliatæ. (Nomen *Pterolepis* in partem tantum specierum quadrat).

21. HYMENOCHÆTE (*P. de B.*)—Perigynii setæ filiformes, molles, pubescentes, deciduæ. Stylus trifidus.

22. ERIOPHORUM (*Linn.*)—Perigynii setæ plurimæ, capillares, longissimæ, basi in fasciculos collectæ, comam construentes. Stylus trifidus.

23. SCIRPUS (*P. de B.*)—Perigynii setæ rigidæ, persistentes. Stylus bi-trifidus, æqualis, deciduus. Caryopsis biconvexa, papillata.

24. ELEOGENUS (*N. ab E.*)—Perigynii setæ rigidæ, persistentes (in una specie nullæ). Stylus bifidus, bulbo baseos calloso in fructu persistente. Caryopsis biconvexa, lævissima, atra. (Spicula solitaria, terminalis).

25. ELEOCHARIS (*R. Br.*)—Perigynii setæ rigidæ, persistentes. Stylus bifidus aut trifidus, bulbo baseos suberoso rugoso in fructu persistente. Caryopsis obovata biconvexa. (Spicula solitaria terminalis).

26. LIMNOCHLOA (*P. de B.*)—Perigynii setæ rigidæ, persistentes. Stylus trifidus, bulbo baseos cartilagineo compresso coloratoque in fructu persistente. (Spicula solitaria, terminalis, squamis pulchre spiralibus rigidulis).

TRIB. IV. RHYNCHOSPOREÆ.

Hermaphroditæ aut sexu in eadem spicula discreto. Spiculæ distichæ, subunifloræ, rarissime pluriflore, squamis inferioribus minoribus sterilibus. Perigynium raro nullum, plerumque setosum persistens aut cupulare. Caryopsis rostrata, regulariter sculpta; rostrum in plerisque discretum.

A. *Stylus bi-trifidus.*

27. RHYNCHOSPORA (*Vahl.*)—Spiculæ polygamæ. Stylus bifidus. Perigynium setosum, setis rigidis denticulatis. Caryopsis styli basi persistente discretaque late rostrata.

B. *Stylus simplex.*

28. MORISIA (*N. ab E.*)—Spiculæ monoecæ. Stylus simplex. Perigynium nullum. Caryopsis rostro brevissimo tuberculiformi aut compresso, a basi styli residuo, coronata.

29. HAPLOSTYLIS (*N. et M.*)—Spiculæ monoecæ. Stylus simplex. Perigynium setosum. Caryopsis acutata aut in rostri formam attenuata, sæpe marginata (subgenus *Pterorhynchium*).

30. CEPHALOSCHŒNUS (*N. ab E.*)—Spiculæ hermaphroditæ. Stylus simplex. Perigynium setosum, setis antrorsum denticulatis aut hirsutis. Caryopsis styli basi persistente discretaque rostrata.

TRIB. V. SCLERIEÆ.

Diclines. Spiculæ fœmineæ uniflorae. Caryopsis nucamentacea, nuda, aut perigynio pateriformi lobato cincta.

31. SCLERIA (*Berg.*)—Nux perigynio lobato aut repando cincta.

32. CYLINDROPUS (*N. ab E.*)—Nux perigynio arcto cylindrico truncato basi constricto.

33. HYPOPORUM (*N. ab E.*)—Perigynium nullum. Nux nuda, basi trigona, lateribus porosis punctatis aut sulcatis.

34. Diplacrum (*R. Br.*)—Perigynium proprium nullum aut disciforme. Nux globosa, squamis binis superioribus persistentibus inclusa.

Trib. VI. ELYNEÆ.

Diclines. Spiculæ compositæ, squamis undique imbricatis, flosculum unum unisexualem aut duos diclines plerumque valvatos tegentibus. Perigynium proprium nullum. Caryopsis plerumque rostrata. Stylus trifidus in speciebus cognitis.

35. Cobresia (*Willd.*)—Spiculæ aggregatæ diclines. Squamulæ propriæ flosculi fœminei binæ, masculi nullæ.

36. Trilepis (*N. ab E.*)—Spiculæ aggregatæ, diclines. Squamulæ propriæ flosculi fœminei et masculi singulæ aut binæ.

Trib. VII. CARICEÆ.

Diclines. Spiculæ undique imbricatæ. Caryopsis tota perigynio inclusa.

37. Carex (*Linn.*)—Caryopsis intra perigynium proprium absque seta.

38. Uncinia (*Pers.*)—Seta hypogyna intra perigynium proprium, sæpe hamata.

Trib. I. CYPEREÆ.

I. CYPERUS. *Linn.*

§ 1. *Stylo bifido, caryopsi compressa.*—Pycreus.

1. C. *pygmæus* (Vahl:) spiculis in capitulum globosum decompositum bracteolis fasciculorum interstinctum dense confertis ovatis compressis 10–14-floris, squamis cymbiformibus acutis membranaceis pallidis dorso 3–5-nervibus, stylo bifido, coryopsi oblonga utrinque acuta convexo-plana subtiliter punctulato-striata, involucri tri-tetraphylli foliolis patentibus reflexisve capitulo longioribus foliisque lineari-acuminatis margine apicem versus scabris, culmo triquetro humili, radice annua.—*Vahl. en.* 2. *p.* 313; *R. et Sch. s. v.* 2. *p.* 174. *n.* 45; *Rottb. gram. n.* 20. *t.* 14. *f.* 4, 5?; *Wight. cat. n.* 1807; *herb. Royle. n.* 100.—[C. diffusus, *Roxb.* (nec *Vahl.*) *fl. Ind.* 1. *p.* 188; (*ed. Wall.*) 1. *p.* 193; *in cœt. merc. Ind. or. mus. tab.* 708.—C. mulen-pulu, *Schult. mant.* 2. *p.* 99.—Arn.]—C. squarrosus, *Roxb. fl. Ind.* 1. *p.* 190; (*ed. Wall.*) 1. *p.* 194.—*Wall. cat. n* 3325.—[*Rheed. Mal.* 12. *t.* 54.—Arn.]——In Peninsula, Ava; *Wallich.*

Omnia congruunt apud Rottboellium, at stylus plerumque est bifidus; variat autem (in exemplis Avanis, Wall. n. 3495) et stylo trifido, licet rarissima. Statura omnino *Dichostylis Michelianæ*, sed spicularum structura longe diversissima.

2. C. *mucronatus* (Linn.:) spiculis aggregatis rariusve solitariis sessilibus ad speciem lateralibus ovato-lanceolatis compressiusculis 15–20-floris, squamis ovatis obtusiusculis arctissime imbricatis albis, stylo bifido, coryopsi obovata plano-convexa, involucri subdiphylli foliolo altero erecto culmum continuante, altero infero umbellam subæquante, culmo teretiusculo adscendente infra spiculas trigono basi monophyllo, folio lineari plano culmo breviori, radice repente geniculata.—*Linn.*; *Roxb.? fl. Ind.* 1. *p.* 185; (*ed. Wall.*) 1. *p.* 190; *Retz. obs.* 5. *p.* 10. *n.* 8; *Link. h. Ber.* 1. *p.* 302. *n.* 51.—Cyperus,

Wall. cat. n. 3311.——Locis humidis scaturiginosis, in variis Peninsulæ Indiæ orientalis regionibus. Ægyptus; *Sieber; Raddi.*

Capitulo spicularum albarum ad speciem laterali, habituque ex eo proficiscente junceo, inter omnes hujus generis sine opera distinguitur. A *C. laterali* differt folio ad radicem culmi sat longo plano, culmo debiliore adscendente nec stricto, involucro superiori incurvo.

[Nullam sub hoc nomine speciem descripsit Linnæus.—Nescio an huc, an species nova sit *C. mucronatus*, Roxb., cujus descriptioni sequentia e tab. picta n. 704 (sub *C. acuminato*) desumpta addentur:—Herba late repens vel potius stolonifera ut in *C. arenario.* Stolonum geniculi squama oblongo-lanceolata brunnea instructi: culmi distantes, adscendentes versus basin squama (ut in stolonibus) muniti, vix ad medium vagina foliifera inclusi. Folium circiter 2½ uncias longum. Squamæ spicularum late ovalia, valde concava, acutiuscula. Caryopsis subrotunda, plano-convexa.—In *C. pleurantho*, radix fere cæspitosa, culmique valde approximati, ut in *C. junciformi*, Desf., *C. distachyo*, Willd., *C. Pannonico*, Linn., quibusdamque aliis, nunc una cum *C. mucronato* tam Retzii tam Vahlii in unicam speciem ab omnibus fere botanicis conjunctis, attamen secundum diagnosin ope exemplorum ex Ægypto provenientium supra datam a cl. Neesio verosimiliter distinguendis. Formam *C. mucronati* typicam egomet ignoro.—ARN.]

3. C. *pleuranthus* (N. ab E.:) spiculis aggregatis ad speciem lateralibus lanceolatis compresso-teretibus multifloris, squamis ovatis obtusis arctissime imbricatis albidis griseove variis, stylo bifido, caryopsi subrotunda planoconvexa, involucri subdiphylli foliolo altero stricto culmum continuante, altero infero spiculas æquante, culmo teretiusculo apice trigono stricto basi vaginato, vagina microphylla, radice repente geniculata.—C. lateralis, *Roxb. fl. Ind.* 1. *p.* 186; (*ed. Wall.*) 1. *p.* 190; *in cat. m. I. o. mus. tab.* 2014 (fide *Arn.*); *Sch. mant.* 2. *p.* 96. [Huc, nec ad C. mucronatum, refero *Wall. cat. n.* 3311.—ARN.]——Coromandel; *Roxburgh.* Insula Mown; *Macrae.*

Differt a *C. mucronato*: culmo magis tereti, vaginæ baseos foliolo minuto filiformi, caryopsi ex ovato subrotunda [magis compressa, acuta, cinerascente subtilissime elevato-punctulata. *C. lateralis*, Forsk., in Ægypto nascens, stylo trifido differens, nescio an speciem propriam exhibeat, quin forsan cum *laterali* Roxb. et *mucronato* in unam speciem debet conjungi. Discrimen nos inter atque illum sic explico:—

* C. *lateralis* (Forsk.:) culmo triquetro, vaginis foliiferis, spiculis geminis ternisve lanceolatis convexis, squamis subrotundis mucronulatis candidis, caryopsi convexo-plana, stylo trifido, involucro diphyllo, altero culmum continuante, altero brevi vix spiculas excedente.—C. mucronatus β albidus, *Vahl.* —C. mucronatus, *Rottb. gr. t.* 8. *f.* 4.—C. leucostachys, *Link. ex herb. Willd.; Link. jahrb.* 2. *p.* 82.——In Ægypto; *Raddi in herb. Endlicher.*

C. pleuranthus seu *lateralis*, Roxb.: " stigmate bifido," ab hoc longe distat.

* C. *macrorrhizus* (N. ab E.:) culmo semitereti compresso hinc sulcato, vaginis longe foliiferis membranaceis, spiculis aggregatis sessilibus oblongis convexis, stylo trifido, squamis ovatis mucronatis carinatis albis basi purpureostriatis, involucro diphyllo capitulum superante, altero culmum continuante prælongo foliisque culmo longioribus complicato-canaliculatis margine serrulato-scabris, fibris radicalibus tomentosis crassis.—Schœnus macrorhizus, *Rich. in herb. Kunth.*——In Ægypto.

Differt a *C. pleurantho* nostro et *laterali* culmo crasso semitereti, foliis culmum superantibus, spiculis duplo majoribus latioribus 18–20-floris, squamis longius mucronatis.

4. C. *angulatus* (N. ab E.:) spiculis approximato-spicatis, spicis singulis geminisve, quandoque et singulis, spiculis ad speciem lateralibus oblongo-lanceolatis compressis 10–12-floris, squamis ovatis argute carinatis acutiusculis cervinis chartaceis, lateribus enervibus, carina trinervi pallidiori, stylo bifido,

caryopsi obovata mucronata plano-convexa punctulato-striata, involucro universali monophyllo erecto culmum continuante foliisque complicato-filiformibus margine scaberrimis, culmo filiformi rigido irregulariter quadrangulari basi folioso.—Cyperus, *Wall. cat. n.* 3324. *a, b.*——In Nepalia et Ava; *Wallich.*

Ex affinitate quidem est *C. mucronati* et *lateralis*, sed spicularum indole et situ longe distat.

5. C. *stramineus* (N. ab E.:) spiculis spicatis imbricatis in spicam simplicem geminasve oligostachyas dispositis lanceolatis utrinque convexulis totis pallide flavescentibus multifloris, squamis ovatis mucronato-acutis lævibus dorso tricarinatis, stylo bifido, caryopsi lenticulari marginata rugulosa, involucri 2–3-phylli foliolis umbella longioribus foliisque lineari-filiformibus apice scabris, culmis incurvis geminis, altero triquetro, altero compresso hinc canaliculato.—Cyperus, *Wall. cat. n.* 3320.——In Silhet; *Wallich.*

6. C. *pumilus* (Linn.:) umbellæ compositæ radiis strictis, spiculis in spicas ovatas confertis reflexo-patentibus oblongis compressiusculis 8–18-floris, squamis ovatis latere pallidis nervosis carina viridibus mucronatis mucrone patulo, stylo bifido, caryopsi obovata compressa retusa punctulata, involucri 3–5-phylli foliolis 2–3 umbella longioribus latiusculis planis caulinisque culmum æquantibus aut superantibus lævibus, culmo basi folioso triquetro lævi, radice fibrosa.—*Linn. amoen. acad.* 4. *p.* 302; *sp. pl. ed.* 2. *p.* 69 (excl. syn. *Pluken.*); *R. et Sch. s. v.* 2. *p.* 188. *n.* 92 (excl. omnibus synon. præter *Linneanum*); *Wight. cat. n.* 1808.—C. membranaceus, *Vahl. en.* 2. *p.* 330; *R. et Sch. l. c.*—Cyperus, *Wall. cat. n.* 3336. *c* ? (ex *Arnott*).——Peninsula Indiæ orientalis; *Wight.*

Verba Linnæi rem acu tangunt: similis *Cypero compresso*, sed brevior; distinctus: glumis minus imbricatis et apice mucronatis; spiculæ vix alternæ, fere digitatæ.—Spiculæ 8–10 in summitate radiorum ita spicatim approximatæ, ut fere capitulum constituant, 3–4 lineas longæ, 1½ lineas latæ, 7–10 floræ; inferiores reflexæ, sæpe geminatæ. Squamæ imbricatæ, mucrone viridi patulo, lateribus late albo-membranaceis nervis ternis incurvis prominulis; carina viridis, nervuloso-striata; mucro validus, viridis, obtusiusculus. Stamina duo. Caryopsis pallide brunnea, compressa, retusa, punctulato-striata. Radii præter umbellulam sessilem 4–5, quorum longior 2-pollicaris. Involucri folium longius 6 poll. longum, duas lineas latum, incurvum. Folium radicale in nostro unum, culmum æquans, paulo angustius; omnia folia lævia plana, aut complicato-carinata. Culmus in nostro haud strictus, basi incurvus, vix pedalis.

[Huic etiam folia tam radicalia quam involucri planiuscula et ad speciem flaccida, et subpatula: radii umbellæ porro sunt patuli. *C. strictus* autem Roxb. (fl. Ind. 1. p. 200; (ed. Wall.) 1. p. 203, vel *C. rigidus*, Roxb. in cœt. merc. angl. Ind. or. mus. tab. pict. 721) differre videtur culmis foliis angustis radicalibus et involucralibus atque umbellæ radiis, rigide ac strictim erectis: discrimen tamen aliud vix ullum indicare possum. *C. pumilo*, Linn., valde affinis est *C. pumilus*, Roxb. (*C. flavescens*, Roxb. in cœt. merc. I. O. mus. tab. 717), nec forsan differt.—ARN.]

7. C. *pulvinatus* (N. et M.:) spiculis globoso-capitatis lanceolatis compressis multifloris pallidis, capitulis solitariis aut aggregatis, squamis late ovatis ex apice obtuso patulo brevimucronatis dorso quinquinervibus, rachilla compressa foveolis bifariam insculpta, stylo bifido, caryopsi obovata truncata compressa utrinque convexula punctulis exasperata, involucri triphylli foliolis binis umbellam superantibus foliisque linearibus angustis culmo longioribus lævibus, culmo humili triquetro.—β; culmis humillimis, pulvinatis.—*N. et M. in Mey. it. ined.*; *Wight. cat.n.* 1809.—Cyperus, *Wall. cat. n.* 3312.—C. punctatus, *Roxb. fl. Ind.* 1. *p.* 193; (*ed Wall.*) 1. *p.* 197 (ex *Wall.* cat. L. n. 3312, et *Arn.* in litt. tab. pictam Roxb. in cœt. merc. Ind. or. mus. n. 712 citante).—Peninsula Ind. or.; *Roxburgh*; *Wight.*

A *C. nitente*, Retzii, differt spiculis neutiquam nitentibus, nec medio fuscis, stylo bifido, nec trifido. A *C. pumilo*, qui etiam stylo est bifido, differt squamis pallidis, nec coloratis, ut habet Gærtnerus (nam Linnæus tacet colorem), neque icon Pluk. t. 191. f. 8. repugnat nostro. *C. punctatus*, Roxb., guadet fructibus albo-punctatis, similisque esse dicitur iconi Pluk. t. 192, f. 3, quæ icon habitu longe distat a nostro.

[Hac tamen certe pertinet *C. punctatus* quamvis per errorem in tabula pictę supra citata, stamina 3 delineantur. Quoad Plukenetii iconem, tab. 192 per errorem citata est pro tab. 191, et pro figura 8 supra a Neesio huc referta, in exemplo meo, forsan etiam in Roxburghiano, 3 excudebatur: Roxburghius itaque revere citat Pluk. t. 191. f. 8; dum t. 192. f. 3 ad *C. difformem*, et recte ut mihi videtur, refertur.—ARN.]

8. C. (P.) *squarrosus* (Rottb.:) spiculis spicato-capitatis linearibus angustis patenti-confertis 10–14-floris, capitulis hemisphæricis in umbellam triradiatam dispositis, squamis ovato-lanceolatis patulo-acuminatis dorso trinervibus viridibus margine membranaceo-albis demum fuscis stamine uno, stylo bifido, caryopsi lanceolata subtrigona utrinque acuta punctato-aspera, involucri 3–5-phylli foliolis plerisque umbella longioribus foliisque culmum triquetrum humilem superantibus linearibus angustis lævibus.—*Rottb. gram. p.* 25. *n.* 29. *t.* 6. *f.* 3; *Vahl. en.* 2. *p.* 331; *Roem. et Sch. s. v.* 2. *p.* 189; *Wight. cat. n.* 1810.—C. Maderaspatanus, *Willd. sp. pl.* 1. *p.* 278 (excl. syn. *Pluk.*).—*Wall. cat. n.* 3312. *d.*——Peninsula Ind. or.

[9. C. *tortuosus* (Roxb.:) umbella composita, spiculis linearibus incurvatis 20–40 floris, squamis ovalibus mucronulatis castaneis, staminibus 2, stylo bifido, caryopsi obovata lenticulari compressa lævi, involucri triphylli umbellam superantis foliolis patulis, foliis subradicalibus culmo obtuse trigono erecto lævi brevioribus patulis carinatis culmi basin spiraliter vaginantibus.—*Roxb. fl. Ind.* 1. *p.* 197; (*ed Wall.*) 1. *p.* 201; *in cæt. merc. Ind. or. mus. tab.* 718; *Wall. cat. n.* 3339.——In pascuis udis Circarensibus; *Roxburgh.*

Habitu omnino refert *C. hexastachyi* varietatem α; at caryopsi compressa, stylo bifido, et staminibus binis differt: hæc tamen discrimina a Roxburghio indigitata erant non ex autopsia sed, ut in aliis quibusdam, ex tabula a pictore forsan quoad hæc haud satis accurate delineata.—ARN.]

10. C. (P.) *sanguinolentus* (Vahl:) umbella composita in capituli formam contracta, spiculis ovato-oblongis 10–12 floris, squamis late ovatis obtusiusculis apice contractis dorso membranaceo pallidis rubro-irroratis trinervibus limbo fusco-sanguineis, stylo bifido, caryopsi lenticulari compressa, involucri 2–4-phylli patentis foliolis duobus foliisque culmum subæquantibus aut brevioribus linearibus planis, culmo humili trigono.—*Vahl. en.* 2. *p.* 351; *Roem. et Sch. s. v.* 2. *p.* 208. *n.* 152.—C. pumilus, *Gærtn. fr. et sem.* 1. *p.* 9. *t.* 2. *f.* 2 (excl. syn.)—β, *major*; umbella magis explicata, radiis 1–4 capituligeris.—C. cruentus, *Roxb. fl. Ind.* 1. *p.* 196; (*ed. Wall.*) 1. *p.* 200 (excl. synon. *Rottb.* et *Willd.*); *Retz. obs.* 5. *p.* 13.—C. pumilus, *Rottb. gram. p.* 29. *t.* 9. *f.* 4 (excl. syn. *Linn.*).—C. ater, *Vahl. en.* 2. *p.* 335; *Roem. et Sch. syst. veg.* 2. *p.* 190. —C. eragrostis *Vahl. en.* 2. *p.* 322; *Roem. et Sch. s. v.* 2. *p.* 180.—Cyperus, *Wall. cat. n.* 3319.—*Pluk. alm. t.* 191. *f.* 8?——In Himalaya var. α legit *Royle.*—β. In Nathpur; *Hamilton.* In Benghala; *Wallich.* In Connahor; *Wight.*

C. cruentus, Rottb., in Arabia nascens, major est styloque gaudet trifido et caryopsi trigona.

11. C. (P.) *polystachyus* (Rottb.:) umbellæ radiis subcorymbosis plerisque abbreviatis confertis involucro tri-pentaphyllo basi serrulato-scabro brevioribus, spiculis lineari-lanceolatis multifloris stramineis, valvis ovalibus obtusis cum mucronulo exiguo lævibus nervo costali viridi bisulco, caryopsi ovali compressa retusa minute punctulata cinerea, foliis linearibus carinatis culmo trigono brevioribus lævibus, stylo bifido, staminibus duobus.—*R. Br. prod. fl. Holl.*; *Link. fl. Ber.* 1. *p.* 310; *Roxb. fl. Ind.* 1. *p.* 193; (*ed Wall.*) 1. *p.* 197;

R. et Sch. s. v. mant. 2. *p.* 102 ; *N. ab E. in Linn.* 7. *p.* 515 ; *Wight. cat. n.* 1811.—C. corymbosus *Roxb. in cæt. merc. I. o. mus. tab.* 711 (fide *Arn.*)—Cyperus, *Wall. cat. n.* 3340. *b* (ex parte, pars altera ad var. γ spectat.).—β ; tenerior, radiis omnibus longiusculis ¼–1 pollicem longis gracilibus, spiculis paulo angustioribus lateribus brunescenti maculo notatis.—γ ; speculis omnibus in capitulum subglobosum congestis.—*Wight. cat. n.* 1811. *b.*——α ; In capite Bonæ Spei, Insula St. Mauritii, Nova Hollandia, Oahu, China, Penins. Ind. or.—β ; In China.—γ ; Ad Rio de Janeiro, et in Pen. Ind. or.

C. polyst., Rottb., stylo trifido describitur atque pingitur ; an errore? Certo etenim reliqua omnia et descriptionis et iconis mirifice congruunt cum exemplis, quæ compararimus.

12. C. *capillaris* (Koen. :) spiculis spicatim capitatis divaricatis lineari-lanceolatis multifloris, umbella 1–3-radiata, squamis ovatis obtusis dorso tricarinulatis lateribus lævigatis chartaceis pallide brunneis, rhachilla compressa lateribus undatim carinatis marginibus foveatis, stylo ad medium fere bifido, caryopsi obovata compresso-biconvexa lævi, involucro subtriphyllo patulo, foliis binis umbella longioribus, foliis subradicalibus geminis anguste linearibus convoluto-subulatis lævibus altero brevissimo, altero culmum compresso-trigonum striatum æquante.—*Koenig. ap. Roxb. fl. Ind.* 1. *p.* 194 ; (*ed Wall.*) 1. *p.* 198 ; *Sch. Mant.* 2. *p.* 103 ; *Royle. herb. cyp. n.* 6. *et* 7 ; *Wight. cat. n.* 1812.—C. capillaceus, *Roxb. in cæt. merc. I. o. mus. tab.* 713 (ex *Arn.*)—Cyperus, *Wall. cat. n.* 3318.——Kamaon, Mysore, Himalaya.

Accedit iste *Cyperus* ad *C. flavescentem* et *polystachyum*, differt autem ob isto spiculis latioribus, ad illo spiculis angustioribus, ab utroque culmo haud evidenter trigono basin versus compresso et involucro rigidulo. Differt a *C. Linneano* et *lanceolato* Poir., herb. Lindl ; spiculis latioribus et rachillæ sculptura.

13. C. (P.) *inundatus* (Roxb. :) spiculis racemosis alternis brevipedicellatis patentibus multifloris, roseo vel luteo alboque variis, racemis in umbellam compositam dispositis, umbellulis late pyramidalibus nudis, squamis late ovatis obtusis imbricatis tenuibus dorso nervis tribus viridibus præditis, latere roseis aut luteis, margine albo membranaceo undulato, rachilla albo-membranacea, stylo bifido, caryopsi obovata retusa compressa punctis minutis asperula, involucri 4–5-phylli foliolis exterioribus umbella longioribus foliisque radicalibus culmum superantibus carinatis lævibus, culmo triquetro angulis lævibus obtusiusculis.—*Roxb. fl. Ind.* 1. *p.* 201 ; (*ed Wall.*) 1. *p.* 205 ; *in cæt. merc. Ind. or. mus. tab.* 1100 (ex *Arn.*) ; *Wight cat. n.* 1813.—C. Bacha, *Ham.*—C. procerus, *Roxb. fl. Ind.* 1. *p.* 203 ; (*ed Wall.*) 1. *p.* 206 (ubi loco " involucres none," legendum " involucels none").—Cyperus, *Wall. cat. n.* 3336 ; *Roxb. in cæt. merc. I. o. mus. tab.* 725 (fide *Arn.*)——In Peninsula Ind. or. et Benghalia.

14. C. (P.) *alopecuroides* (Rottb. :) umbella decomposita pluriradiata radiis compressis strictis apice fasciculatim-umbellatis e spicis pluribus patulis recurvisve cylindricis a basi dense imbricatis, spiculis ovato-oblongis 10–14-floris compressis ; squamis ovatis mucronulato-acutatis dorso viridi nervosis margine flavis incurvisque, stylo bifido, caryopsi ovata convexo-concava lævi, involucellis e basi vaginante herbacea linearibus umbellula brevioribus, involucri 6–7-phylli foliolis 3–4 late linearibus umbella longioribus foliisque margine scabris, culmo trigono.—*Rottb. gram. p.* 38. *t.* 8. *f.* 2 ; *Vahl. en.* 2. *p.* 368 ; *R. et Sch. s. v.* 2. *p.* 220 (excl. omnibus synonymis) ; *N. ab E. in Cyp. Royl. n.* 11.——In Himalaya ; *Royle.*

§ 2. *Caryopis trigona, stylo bi-trifidove. Spiculæ valde compressæ, squamis acutis acute carinatis dorso herbaceo-striatis.*—Compressi, *Lindl.*

17. C. *compressus* (Linn. :) umbellæ compositæ simplicisve radiis gracilibus patulo-erectis spiculis capitato- aut digitato-spicatis in disco umbellæ

aggregatis lanceolatis ancipitibus multifloris viridulis pallidis, squamis ovatis carinatis acuminatis imbricatis purpura lineolatis latereve luteo-notatis dorso tenuinervibus, stylo trifido, caryopsi obovata trigona pallide fusco-cinerascente basi truncata area detrita, involucri 3–4-phylli foliis plerisque umbella longioribus foliisque linearibus attenuatis apicem versus serrulato-scabris, culmo trigono.—*Linn. sp. pl. p.* 68; *Willd. sp. pl.* 1. *p.* 282. *n.* 49; *Vahl. en.* 2. *p.* 324; *R. et Sch. s. v.* 2. *p.* 182; *mant.* 2. *p.* 106; *Rottb. gram. p.* 27. *t.* 9. *f.* 3; *Roxb. fl. Ind.* 1. *p.* 194; (*ed. Wall.*) 1. *p.* 198 (excl. syn. *Pluk.*); *H. et K. nov. gen. et sp. pl.* 1. *p.* 167; *Link. en.* 1. *p.* 45; *H. Ber.* 1. *p.* 309; *Wight. cat. n.* 1814—C. viridis, *Roxb. in cœt. merc. Ind. or. mus. tab.* 714 (fide *Arn.*)—C. conglomeratus, *Willd. en. suppl. p.* 5. (*nec sp. pl.*)—Cyperus, *Wall. cat. n.* 3308.—*Sloane Jam.* 1. *t.* 76. *f.* 1; *Pluk. t.* 417. *f.* 2.——India utraque. Peninsula Indiæ orientalis, solo humido.

C. brachiatus, Poir., et *pectinatus*, Roxb., ut subspecies huic et nostro subjungendæ sunt. Differt autem: *C. brachiatus* radiis umbellæ fortioribus ad horizontem patentibus, spiculis latioribus, coryopsi breviori latiori atra nec cinerea; *C. pectinatus* culmo humiliore squamisque longius acuminatis, spiculis capitatis.

18. C. *pectiniformis* (R. et Sch.:) spiculis aggregatis capitatis divergentibus compressis 25–36-floris pallidis, squamis ovatis patulo-acuminatis carinatis, stylo trifido, caryopsi obovata trigona cinerea lævi, involucro triphyllo spiculas superante foliisque culmum trigonum humilem subæquantibus linearibus angustis, radice fibrosa.—*R. et Sch. mant.* 2. *p.* 128 (loco alieno).—C. pectinatus, *Roxb. fl. Ind.* 1. *p.* 190; (*ed. Wall.*) 1. *p.* 195.—C. Coromandelinus, *Spr. s. v.* 1. *p.* 217.—C. Meyenii, *Wight et Arn.*—β, *subacaulis*; culmo brevissimo, spiculis longissimis.—Cyperus Meynii, *Wight. cat. n.* 1815.——In Coromandel; *Roxb.* In Manila insula; *Meyen.* In China; *Vachell.*—β; In Peninsula Ind. or.; *Wight.*

Quod ad spiculas, hæc species est ex affinitate *C. compressi* (cujus probabiliter est varietas), *brachiati*, &c. In var. β culmus vix semipollicaris spiculas profert 4–6, ultra duos pollices longas, curvulas, 70–80-floras. Reliqua omnia congruunt cum forma communi. Macula flava squamarum distincta.

[Diagnosis supra data vix ac ne vix ad plantam Roxburghianam seu *C. pectiniformem* spectat. In *C. pectiniformi* β, N. ab E., iisdem exemplis mihi nunc ante oculos jacentibus, spiculæ minores unciam longitudine excedunt, at majores et fere omnes duas pollices longe evadunt: squamæ valde numerosæ, utrinque 20–36, spiculæ igitur sæpius flores plus quam 60 exhibent. In *C. pectinato*, Roxb., tamen, spiculæ multo breviores et ad speciem paucifloræ: majores, secundum tabulam suam pictam No. 215, in cœt. merc. Ang. Ind. or. museo asservatam, 9–10 lineas tantum longæ sunt, et vix unquam utrinque 14–15 floribus guadent. In planta Neesiana porro spiculæ revere lineares sunt, et in tabula Roxburghiana lineari-oblongæ aut oblongo-lanceolatæ; in illa culmi vix unciam, in hac 10–12 uncias longi sunt. Allatis omnibus haud dubito quin *C. pectiniformis* verus sit *C. Wightii*, N. ab E. Nec obstat culmi in exemplo unico Wightiano, casu tantum forsan, gracilior et flaccidor.—ARN.]

§ 3. *Squamæ membranaceæ, nervoso-striatæ; caryopsis trigona, stylo trifido.*— Hymenolepides, *Lindl.*

17. C. *arenarius* (Retz.:) spiculis subgloboso-aggregatis ovatis convexis sudecemfloris, squamis imbricatis acutiusculis (infimis obtusis) membranaceis tenuiter nervulosis pallide ferrugineis, stylo trifido, caryopsi trigona, involucri 2–3-phylli foliolis uno binisve elongatis foliisque semitereti-filiformibus canaliculatis lævibus, culmo teretiusculo subtrigono basi folioso, radice longe repente sarmentosa.—*Retz. obs.* 4. *p.* 9. *n.* 11; *Willd. sp. pl.* 1. *p.* 269; *Vahl. en.* 2. *p.* 298; *R. et Sch. s. v.* 2. *p.* 161. *n.* 2; *Wight. cat. n.* 1816.— Cyperus, *Wall. cat. n.* 3535.—*Pluk. alm. t.* 300. *f.* 7.——In solo arenoso ad litus maris.

18. C. *Wightii* (N. ab E. :) spiculis globoso-capitatis oblongo-lanceolatis compressis 10–30-floris, squamis arcte imbricatis ovatis in acumen mediocre subulatum recurvum attenuatis subtiliter nervoso-striatis pallide brunneis margine albo-membranaceo enervi, stylo trifido, caryopsi obovata trigona læviuscula cinerea, involucri di-triphylli foliolo altero capitulo subduplo longiori, foliis anguste linearibus complicato-canaliculatis culmo flaccido triquetro brevioribus, radice fibrosa.—C. pectiniformis, *Wight. cat. n.* 1817.——Peninsula Indiæ orientalis ; *Wight.*

Solus huic affinis est *C. Teneriffæ*, Poir. ; differt autem hic : foliis longioribus, spiculis majoribus rufis, squamis evidentius striatis magisque acuminatis. Inter Indicas species hujus generis comparandæ sunt : *C. capitatus*, Retz., fasciculo in culmo foliorum medio insignis foliisque culmo longioribus ; tum vero *C. nitens*, Retz., cui spiculæ monandræ, alia ut taceam. *C. pectinatus*, Roxb., differt culmo stricto et spiculis linearibus : an ad hunc nostrum referendus ?

[Videas *C. pectiniformem.* Quoad *C. capitatum*, Retzii, culmus fasciculum foliorum proferre ad medium tantum fortuito videtur.—*C. nitens*, Vahl. et forsan etiam Retzii, ad *C. castaneum*, Willd., potius spectat.—ARN.]

21. C. *niveus* (Retz. :) spiculis 4–10 capitatim-aggregatis ovato-oblongis subvigintifloris (20–30-floris), squamis imbricatis apice biemarginatis membranaceis nervoso-striatis albidis cervinisve, stylo trifido, caryopsi trigona nigra, involucri 1–3-phylli foliolo altero elongato foliisque anguste linearibus longis lævibus, culmo præter basin aphyllo trigono striato, rhizomate dense multicipite.—*Retz. observ.* 5. *p.* 12 ; *Willd. sp. pl.* 1. *p.* 273 ; *Vahl. en.* 2. *p.* 308 ; *R. et Sch. s. v.* 2. *p.* 169 ; *Roxb. fl. Ind.* 1. *p.* 191 ; *(ed. Wall.)* 1. *p.* 195 ; *in cœt. merc. Ind. or. mus. tab.* 710 (fide *Arn.*) ; *Wight. cat. n.* 1818 ; *Royle. Cyp. n.* 8.—Cyperus, *Wall. cat. n.* 3377.—*β, cervinus ;* culmis altioribus (pedalibus et longioribus), spiculis pallide cervinis.—*N. ab E. in herb. Royle.*—Cyperus, *Wall. l. c.*——*α* ; In Peninsula boreali prope oppidum Rajamundri. In nemorosis ad Midnapur. Tjandracona.—*β* in ripa Irawaddi, in Sirmoore, Kamaon, in Monghir, et Dimla et Sanashygota.

Retzius, ut solet, bene descripsit. Folia perhibens brevia, aperto falsus est, quippe quæ culmum haud raro superent, linearia, rigidula, undatim flexuosa, siccando se convolventia. " Acutas " dixisse videtur squamas Willdenovius, quoniam a tergo eas, neque a latere, neque explanatas observavit.

§ 4. *Caryopsis trigona stylo trifido. Spiculæ lineares squamis carinatis mucrone longo recurvo.*—Aristati, *Lindl.*

22. C. *aristatus* (Rottb. :) spiculis subgloboso-capitatis linearibus seu 14-floris e viridi rufescentibus, capitulis umbellatis aut aggregato-congestis, squamis ovatis in acumen subulatum recurvum attenuatis subtiliter septemnervibus, stigmatibus ternis, caryopsi ovali trigona punctato-striata (styli basi persistente mucronata), involucri tri-tetraphylli foliolis binis capitulo longioribus foliisque anguste linearibus lævibus, his culmum triquetrum æquantibus. *Rottb. gram. p.* 23. *t.* 6. *f.* 1 ; *Vahl. en.* 11. *p.* 369 ; *Willd. sp. pl.* 1. *p.* 275 ; *R. et Sch. s. v.* 2. *p.* 221 ; *Roxb. fl. Ind.* 1. *p.* 190 ; *(ed. Wall.)* 1. *p.* 194 ; *Schult. mant.* 2. *p.* 128 ; *Royle. herb. n.* 26. *et* 27 ; *Wight. cat. n.* 1819.—C. intricatus, *Roxb. in cœt. merc. Ind. or. mus. tab.* 709 (fide *Arn.*)—Cyperus, *Wall. L. n.* 3375.——In pascuis humidis Peninsulæ Ind. or.

Differt a *Cypero versicolore* spiculis duplo minoribus, squamis tenerioribus, e viridi demum subrufescentibus, acumine breviori instructis, nervis haud ita conspicuis striatis. Stamen unum.

23. C. *versicolor* (N. ab E. :) spiculis in capitulum ovatum congestis oblongo-lanceolatis compressis subdecemfloris primum pallide virentibus dein fusco-sanguineis, squamis e basi late ovata in acumen subulatum patenti-recurvum longum attenuatis crasse septemnervibus, stigmatibus ternis, cary-

opsi oblonga trigona punctulato-subundata cinerea, involucri triphylli foliolis duobus elongatis foliisque latiuscule linearibus culmum humilem superantibus lævibus, culmo acute triquetro crasso.—*Wight. cat. n.* 1820.—Cyperus, *Wall. cat. n.* 3374.——Crescit locis arenosis provinciæ Tanjore ad Negapatam; *Wight.*

Similis *Cypero aristato,* sed differt: spiculis constanter capitatis, duplo majoribus, ætate vivide castaneo-purpureis (capitulo phleoideo); squamis costatis, culmo crassiori 1–3 poll. longo; vaginis laxis, purpureis; foliis latioribus.—Stamen unum (in nostris). Caryopsis, pro spicularum et valvularum mole exigna. Involucri folia patentia; tertium vix capitulo longius.

22. C. *angustifolius* (Hamilt.:) spiculis subgloboso-capitatis aut subdigitatis lineari-lanceolatis 5–10-floris, capitulis in umbellam triradiatam dispositis aut subcongestis, squamis ovatis infra apicem obtusum retusumve in acumen subulatum recurvum productis dorso nervoso-tricarinatis (viridulis, margine rufis aut rufo-irroratis), stylo trifido, caryopsi obovata trigona lævi, involucri pentaphylli foliolis umbella longioribus foliisque culmum superantibus convoluto-filiformibus lævibus, culmo setaceo subquadrangulo.—*Hamilt. in Wall. cat. n.* 3376, *b.*——Bengalia inferiore; *Wallich*; *Hamilton.*

Species distinctissima, ante omnes similis *Cypero castaneo,* Willd., qui quidem differt: culmo acutangulo, nec subtetragono (lateribus duobus planinsculis, duobus sulciformibus), spiculis parvis, linea vix longioribus, nec unguicularibus.

23. C. *castaneus* (Willd.:) spiculis digitatis linearibus multifloris in umbella simplici 3–6-radiata, squamis ovatis infra apicem obtusum in acumen subulatum recte patulum productis dorso nervoso-tricarinatis viridibus latere castaneis, stylo trifido, caryopsi ovali trigona punctulato-rugosa, involucri subpentaphylli foliolis umbellam æquantibus superantibusve foliisque anguste linearibus convoluto-canaliculatis subulatis lævibus, culmo humili trigono.—*Willd. sp. pl.* 1. *p.* 278; *Vahl. en.* 2. *p.* 317; *R. et Sch. s. v.* 2. *p.* 177; *Roxb. fl. Ind.* 1. *p.* 195; (*ed. Wall.*) 1. *p.* 199; *in cœt. merc. I. or. mus. tab.* 715 (ex *Arn.*); *Wight. cat. n.* 1821.—Cyperus, *Wall. cat. n.* 3323.——In pascuis arenosis humidis Peninsulæ Indiæ orient.

A *Cypero angustifolio* differt acumine squamarum breviori recto, nec recurvo, $\frac{1}{3}$–$\frac{1}{2}$ longitudinis squamæ, tum colore læte castaneo squamarum. Stylus non bifidus est, ut apud Roxburghium legimus, sed trifidus.

[Roxburghius tamen recte caryopsin trigonam adscripsit.—Arn.]

§ 5. *Caryopsis trigona, stylo trifido. Spiculæ breviusculæ, capitatim congestæ, capitulis umbellatis aut aggregatis subsessilibus. Squamæ acutæ margine membranaceæ.*—Capituligeri, *N. ab E.*

24. C. *Silletensis* (N. ab E.:) spiculis ovatis compressis 10–12-floris capitatis capitulis in umbellam decompositam et supradecompositam densam et brevem dispositis, squamis pallidis oblongis obtusis rectis patulis latere uninervibus nervo tumidulo, stylo trifido, caryopsi ovali triquetra acuta lævi, involucri 6–7-phylli foliolis umbella longioribus erectis foliisque linearibus acuminatis lævibus, culmo triquetro lævi basi folioso.—Cyperus, *Wall. cat. n.* 3363, *f.*——E. Sillet.

25. C. *pallidus* (Heyne?:) umbella supradecomposita depressa, radiis alternis brevibus, spiculis in apice radiolorum 4–6 contiguis subpalmatim patentibus oblongo-lanceolatis compressis 10–14-floris, squamis arcte incumbentibus ovatis mucronulatis carina tricostata lateribus ex luteo pallescentibus, staminibus ternis, stylo trifido, caryopsi ovali trigona, involucellis 2–3-phyllis linearibus ciliatis, involucro hexaphyllo umbella longiore, foliolis foliisque elongatis latiusculis planis margine carinaque scaberrimis glaucis, culmo triquetro lævi.—Cyperus, *Wall. cat. n.* 3359, *d.*——In Penang; *Wallich.*

Similis est *Cypero vegeto,* a quo differt colore pallide glauco, spiculis minoribus e canescente luteis magis compressis bifariam patentibus et veluti flabella parva in apicibus radiolorum exhibentibus, staminibus ternis, nec uno tantum.

26. C. *Haspan* (Rottb.:) umbella supradecomposita laxa, spiculis linearibus compressis in cymulas divaricatas dispositis, squamis ovatis obtusatis mucronatisque triandris, latere unisulcatis, stylo trifido, caryopsi exigua globoso-trigona alba, involucro universali diphyllo umbella breviori aut altero folio longiori, partialibus bipaleaceis, culmo acute triquetro basi vaginato monophyllove.—α; corymbulis pleiostachyis.—*Wight. cat. n.* 1822.—C. Haspan, *Rottb. gram. p.* 36. *t.* 6. *f.* 2; *Willd. sp. pl.* 1. *p.* 287. *n.* 69; *Vahl. en.* 2. *p.* 320; *R. et Sch. s. v.* 2. *p.* 179; *Roxb. fl. Ind.* 1. *p.* 210; (*ed Wall.*) *p.* 213; *in cœt. merc. Ind. or. mus. tab.* 734 (fide *Arn.*).—Cyperus, *Wall. cat. n.* 3369.—C. strictus, *Heyne in Wall. cat. n.* 3365, *c.*—β; cymulis bifidis cum spicula intermedia sessili, ramis divaricatis sæpe monostachyis.—Cyperus, *Wall. cat. n.* 3369. *e.*—γ; autumnalis, culmo laxiori siccando valde collabescente, foliis caulinis nullis, involucri diphylli folio altero erecto elongato, umbella composita decompositaque radiis flaccidioribus recurvo-patentibus*.—*Wight. cat. n.* 1822, *b.*—C. autumnalis, *Vahl. en.* 2. *p.* 318 (excl. C. autumnali, *Rottb.?*)—C. aphyllus, *Rohr. in Vahl. ecl. Am.* 2. *p.* 6.—Cyperus, *Wall. cat. n.* 3368.——α; In montibus Peru prope Tarma; *Meyen.* In Tavoy; *Wallich.*—β; In Asia et America tropica et alpina.—γ; In Ceylona, in Penins. Ind. or., Silhet, et in America meridionali.

27. C. *articulatus* (Linn.:) umbellæ supradecompositæ laxæ spiculis subcapitatis linearibus compressis, stylo trifido, caryopsi trigona, involucro triphyllo brevi, culmo aphyllo nodoso.—α; culmo graciliore internodiis longioribus, umbellæ radiis oligostachyis.—C. articulatus, *Vahl. en.* 2. *p.* 301; *Willd. sp. pl.* 1. *p.* 270; *Link. h. Ber.* 1. *p.* 301; *H. et K. nov. gen. et sp. p.* 184.—β, *Niloticus;* umbellæ densæ radiis polystachyis, culmo crasso, articulis brevibus.—*Wight. cat. n.* 1823.—C. Niloticus, *Forsk. descr. p.* 13; *Vahl. en.* 2. *p.* 302; *R. et Sch. s. v.* 2. *p.* 164.—C. nudus, *Roxb. fl. Ind.* 1. *p.* 187; (*ed Wall.*) 1. *p.* 191; *in cœt. merc. I. o. mus. tab.* 705 (ex *Arn.*).—C. gymnos, *R. et. Sch. mant.* 2. *p.* 97; *Willd. sp. pl.* (*ed. Dietr.*) 1. *p.* 208. *n.* 2.—Cyperus, *Wall. cat. n.* 3364.——α; In rivulis Caribæorum et Guinea, in planis calidis littorum Oceani Pacifici et prope Bordones Cumanensium, ad Arica et Tarma Peruviæ; in Insula St. Mauritii, et Peninsula Indiæ orientalis.—β; In aquis stagnantibus aut lente fluentibus Peninsulæ Indiæ orientalis; *Roxburgh, Wight.*

A *C. articulato,* si demis characteres supra memoratos, var. β haud differt.

28. C. *seminudus* (Roxb.:) culmo continuo tri-sexpedali tereti superne subangulato aphyllo basi vaginato, umbella decomposita, involucri triphylli foliolis umbella duplo brevioribus margine carinaque scabris, stylo trifido, caryopsi trigona.—*Roxb. fl. Ind.* 1. *p.* 187; (*ed Wall.*) 1. *p.* 191.—C. Koenigii, *Vahl. en.* 2. *p.* 302 (fide *Arn.*)——Locis humidis Indiæ orientalis.

Hunc ego non vidi.

[Secundum tab. Roxburghii pictam, n. 706, spiculæ in radiis alternæ, minimeque capitatæ. Forsan ad *Papyrum* potius pertinet, haud procul a *P. Pangorei* et *tegetiformi* locanda species.—ARN.]

§ 6. *Caryopsis trigona, stylo trifido. Spiculæ lineares aut lanceolatæ in radiis alternæ umbellam rarius contractam formantes. Squamæ dorso striatæ, margine læves.*—Allagostachyi, *Lindl.*

29. C. *bulbosus* (Vahl:) spiculis lineari-lanceolatis acuminatis 10–16-floris

* Huic varietati affinis est *C. flavidus,* Roxb. (vix Retz); at an revere huc, annon, vix quamquam ope tab. pictæ in cœt. merc. I. o. mus. n. 722 (sub nom. *C. trichotomi*) satis certus sum: umbellæ radii multo strictiores sunt. Quomodo autem se habet, in diagnosi Roxburghiana pro "generally shorter" legendum "much longer" ut facile patet ex descriptione et tabula picta.

in apice culmi alternis, infimis geminis, squamis ovato-lanceolatis acuminatis nervoso-striatis brunneis carina viridi, stylo trifido, caryopsi oblonga trigona cinerea fusco-punctata, involucri foliolis alternis infimis duobus spica longioribus, foliis filiformibus omnibus radicalibus longe vaginatis, culmo 1–3-pollicari semitereti-trigono, radice bulbosa tunicata fibris bulbiferis.—*Vahl. en.* 2. *p.* 342 ; *R. et Sch. s. v.* 2. p. 197. *n.* 123 ; *Wight. cat. n.* 1824.—C. jemenicus, *Roxb. fl. Ind.* 1. *p.* 191 ; (*ed Wall.*) 1. *p.* 195 (excl. syn. *Rottb.*) ; *Retz. obs.* 4. *p.* 11.—C. oleraceus *Roxb. in cœt. merc. I. o. mus. tab.* 733 (fide *Arn.*) —C. capitatus, *Retz. obs.* 4. *p.* 9? (planta bulbo exuta).—Cyperus, *Wall.! cat. n.* 3317. *a, b.*——In pascuis siccis arenosisque maritimis Peninsulæ Indiæ orientalis ; *Roxburgh ; Wight ; Koenig.* In Ceylona ; *Rottb.*

An hujus loci : O. capitatus, *Retz. obs.* 4. *p.* 9, " culmo subanguloso striato medio folioso, capitulo terminali triphyllo, spicis sessilibus ?" C. jemenicus, *Rottb.*, quoad iconem et descriptionem diversus omnino a nostro, et *stolonifero* propinquior.

30. C. *stoloniferus* (Retz. :) umbella composita pauciradiata brevi, spiculis 3–7-nis approximato-spicatis subdigitatis oblongo-lanceolatis patulis 10–24-floris compressiusculis, squamis late ovatis ecarinatis obtusiusculis castaneis margine angusto albo-membranaceo undique cinctis uninervibus arcte imbricatis, stylo trifido, caryopsi ovali depresso-trigona lævissima lutea vel brunnea, involucri subtriphylli foliolo altero umbella longiori foliisque culmo brevioribus complicatis angustis margine scabriusculis, involucellis nullis, culmo tenui triquetro, radice repente hinc inde nodosa.—*Retz. obs.* 4. *p.* 10 (descr. opt.) ; *Willd. sp. pl.* 1. *p.* 282 ; *Vahl. en.* 2. *p.* 338 ; *R. et Sch. s. v.* 2. *p.* 194 ; *Wight. cat. n.* 1825.—C. spadiceus, *Lam. ill. gen.* 1. *p.* 147.—C. juncifolius, *Willd. apud Rottb.*—Cyperus, *Wall. cat. n.* 3315.—β ; spiculis brevioribus et pallidioribus.—*Wight. cat. n.* 1825, *b.*——Peninsula Indiæ or. ; *Konig. ; Wight.* Ins. S. Mauritii ; *Sieber.*

31. C. *hexastachyus* (Rottb. :) umbella composita decompositaque tri-quadriradiata, radiis strictis, spiculis spicatim approximatis subsenis (6–10) linearibus convexiusculis 10-multifloris, squamis ovatis obtusis cum mucronulo infraapicali dorso nervoso-striatis purpureis pallescentibusve carina viridi, margine pallido, stylo trifido, caryopsi obovata triquetra punctulata, involucro triphyllo, altero binisve umbella longioribus foliisque linearibus acutis glaucescentibus, his culmo triquetro lævi brevioribus longioribusve, radice repente tuberosa.

A : Communis, foliis culmo brevioribus.—α ; umbella composita.—α*, *communis ;* spiculis rectis 12–20-floris, squamarum margine pallido latiori.—*Wight. cat. n.* 1826.—Cyperus hexastachyus, *Spr. s. v.* 1. *p.* 223 ; *Rottb. gram. p.* 28. *t.* 14. *f.* 2 ; *R. Br. prod.* 1. *p.* 216 ; *R. et Sch. s. v.* 2. *p.* 198 ; *Link. hort. Berl.* 1. *p.* 315.—C. rotundus, *Linn. s. v. p.* 98 ; *fl. Zeyl. p.* 36 ; *Roxb. fl. Ind.* 1. *p.* 197 ; (*ed. Wall.*) 1. *p.* 201. *n.* 24 : *in cœt. merc. I. o. mus. tab.* 719 (ex *Arn.*)—Cyperus, *Wall. cat. n.* 3329, *c, f.*—*Rumph. Amb.* 6. *p.* 1. *t.* 1. *f.* 1 ; *p.* 2. *t.* 1. *f.* 2. —α**, *castaneus ;* spiculis totis fusco-castaneis nitidis.—Cyperus, *Wall. cat. n.* 3322, *e.*—α***, *flexuosus ;* spiculis prælongis 40–50-floris varie curvatis, lateribus squamarum rigidioribus, margine membranaceo angustiori.—*Wight. cat. n.* 1826.—β ; umbella decomposita involucro triphyllo umbella longiore.—β*, *tuberosus ;* spiculis paulo angustioribus purpurascentibus.—*Wight. cat. n.* 1827.—C. tuberosus, *Roxb. fl. Ind.* 1. *p.* 199 ; (*ed. Wall.*) 1. *p.* 203 ; *Rottb. gr. p.* 28. *t.* 7. *f.* 1 (exemplum juvenile) ; *Willd. sp. pl.* 1. *p.* 281 ; *R. et Sch. s. v.* 2. *p.* 196 (excl. syn. *Purshii*).—C. jemenicus, *Roxb. in cœt. merc. I. o. mus. tab.* 720 (ex *Arn.*)—Cyperus, *Wall. cat. n.* 3329, *b.*—[C. procerus *var.* Benghalensis, *Roxb. fl. Ind.* I. *p.* 203 ; (*ed. Wall.*) 1. *p.* 206.—C. tenuiflorus, *Roxb. in cœt. merc. Ang. Ind. or. mus. tab.* 1109.—ARN.]—

β**; spiculis paulo latioribus pallidis distinctius nervosis.—Cyperus, *Wall. cat. n.* 3356.

B: Pendulus, foliis culmo longioribus, fibris radicalibus longis filiformibus apice bulbiferis.—*Wight. cat. n.* 1828.—C. tuberosus, *Roxb. fl. Ind.* 1. *p.* 199; (*ed. Wall.*) 1. *p.* 203 (*excl. syn.*)—Cyperus, *Wall. cat. n.* 3317, *b*, *c*?——Arabia, Indiæ orient. Peninsula et Kamaon, Zeylona, Java, China, Manilla, Nova Hollandia.

32. C. *procerulus* (N. ab E.:) umbella decomposita contracta, radiis valde inæqualibus majoribus trifidis, spiculis spicatim approximatis patulis lanceolato-linearibus sæpe incurvis subtrigintifloris compressis, squamis latiusculis ovatis acutiusculis muticis dorso nervosis fuscoque lineolatis latere e luteo expallescentibus margine omni membranaceis, bracteis loco involucellorum alternis e basi ovata membranacea linearibus ochreolas excedentibus, stylo trifido, caryopsi brevi obovata trigona mucronata punctato-rugulosa, involucri 3–5-phylli foliolis 1–2 umbella longioribus foliisque culmum æquantibus linearibus planis flaccidis margine scabris, culmo basi folioso, trigono lævi, radice bulbosa.—*Wight. cat. n.* 1329.——Peninsula Indiæ orientalis; *Wight.*

Species haud sine difficultate a confinibus *C. hexastachyo* A β, *tuberosa*, *C. Retzii* et ipso *C. compresso* distinguenda, sed mea tamen sententia satis certa. Differt a *C. hexastachyo tuberoso* spiculis latioribus pollicem longis, lineam 1¼ latis, squamis majoribus laxiuscule imbricatis ætate magis patulis haud mucronatis; a *C. Retzii* squamis majoribus haud acuminatis pallidioribus, radice bulbosa, foliis scabris, &c.; a *C. compresso* differt ramis umbellæ tenerioribus, spiculis longioribus, incurvis nec strictis, squamis haud acuminatis.—Sub *C. procero* suo speciminum graciliorum meminit Rottboellius, sibi ex India orientali allatorum, quæ nescio an hujus loci esse existimem, cum in multis icon Retziana, t. 5. f. 3. magis in hunc, quam in *C. procerum* Rothii supra descriptum, (longius a nostro distantem speciem), congruere videatur.

33. C. *Retzii* (N. ab E.:) umbella decomposita supradecompositaque patula, radiis majoribus apice trifidis, spiculis spicatis approximatis alternis patentibus lanceolato-linearibus subvigintifloris compressis, squamis ovatis apice attenuatis acutis muticis ferrugineis dorso nervosis carina viridi, bracteis loco involucellorum alternis e basi latiuscula setaceis ochreolas excedentibus, stylo trifido, caryopsi ovali-trigona, involucri triphylli folio uno alterove umbella longioribus foliisque linearibus planiusculis lævibus, culmo triquetro lævi basi folioso.—*Wight. cat. n.* 1830.—C. Pangorei, *Koenig. in Retz. obs.* 4. *p.* 10; *Willd. sp. pl.* 2. 1. *p.* 284 (excl. syn. *Rottb.*)——Peninsula Indiæ orientalis; *Wight.*

C. procerus, Rottbœllii, proxima species, differt, alia ut taceam, culmo lævi. Descriptio Retziana, l. c. edita ad verbum congruit.

34. C. *tenuiflorus* (Rottb.:) umbella decomposita tri-quinqueradiata radiis capillaribus, spiculis spicatim approximatis subdigitatis (3–5) anguste linearibus acutis convexis 10–20-floris, squamis arcte imbricatis ovatis obtusis muticis, dorso anguste viridi-trinervibus, lateribus lævibus castaneo-brunneis, margine tenuissimo albo, stylo trifido, caryopsi ovali trigona lævi, involucri di-tetraphylli foliolo uno alterove elongato foliisque linearibus carinatis læviusculis, culmo gracili triquetro lævi, radice repente.—*Rottb. gram. p.* 30. *t.* 14. *f.* 1; *Jacq. ic. rar.* 2. *t.* 296; *N. ab E. in cyp. Ecklon. ined.*; *Willd. sp. pl.* 1. *p.* 284; *Vahl. en.* 2. *p.* 347; *R. et Sch. s. v.* 2. *p.* 205; *Link. h. Ber.* 1. *p.* 314 (excl. nb. synon. *Roxb.*); *Wight. cat. n.* 1331.—C. badius, *Desf. fl. Atl.* 1. *p.* 45. *t.* 7. *f.* 2; *Vahl. en.* 2. *p.* 347; *R. et Sch. s. v.* 2. *p.* 203; *Willd. sp. pl.* (*ed. Dietr.*) 1. 2. *p.* 294.—C. incurvatus, *Roxb. fl. Ind.* 1. *p.* 196; (*ed. Wall.*) 1. *p.* 200.—β, *brachystachyus*; spiculis brevioribus 6–7-floris lanceolatis, radiis plerisque apice simpliciter 4–6-stachyis uno autem trifido.—*N. ab E. in herb. Royle, n.* 18.—(*, *pallens*; *N. ab E. l. c. n.* 10.)——*; Peninsula Ind. orient.—β; in Nepalia; *Royle.*

Cum *C. hexastachyo* confundi nequit, propter spiculas multo angustiores castaneo-brunneas magisque attenuatas, squamas margine parum membranaceas, infra apicem (sæpe pallentem) muticas, pauciores plerumque in spicula, umbellæ radios graciliores flaccidos, rhachi inter spiculas tenui flexuosa.

[*C. incurvatus*, Roxb. (in cœt. merc. Ind. or. mus. tab. 2016), vix hujus loci est: huic culmus validus, 1½ lineas crassus, incurvatus, acute trigonus, lateribus concavis; involucri foliola valde carinata; caryopsis lineari-oblonga.—ARN.]

35. C. *pertenuis* (Roxb.:) culmo tri-quadripedali inferne subtereti superne trigono, foliis paucis radicalibus culmo brevioribus umbella composita decompositaque quam pro plantæ altitudine minori, involucro umbellam subæquante, spiculis lanceolatis multifloris, stylo trifido, caryopsi trigona.—*Roxb. fl. Ind.* 1. *p.* 198; (*ed Wall.*) 1. *p.* 202; *in cœt. merc. I. o. mus. tab.* 1320 (fide *Arnott*).——Locis depressis udis in agro Calcuttensi.

[Descriptioni Roxburghiano addes ex tabula sua picta n. 1320: spiculas revere digitales, squamasque subacuminatas et acutas, et culmum basi pennæ anserinæ crassitie.—ARN.]

36. C. *procerus* (Roth:) umbella decomposita compositave pauciradiata subcongestave, spiculis subsenis spicatim approximatis lanceolatis patentibus 20–30-floris compressis, squamis late ovatis obtusis undique albo-marginatis obsolete nervulosis pallide fuscis solo nervo carinali viridi minime excurrente, stylo trifido, caryopsi obovata triquetra lævi, involucri triphylli foliolo uno umbella multo longiore, involucellis nullis, foliis complicato-carinatis canaliculatis margine carinaque scabris, culmo acute triquetro angulis superne scabris.—*Roth. cat.* 3. *p.* 5; *Roxb. in cœt. merc. I. o. mus. tab.* 724 (ex *Arn.*, et itaque forsan, *Wall. cat. n.* 3329. *a*; *Rottb. gram. p.* 29. *t.* 5. *f.* 3 ? ; *Wight. cat. n.* 1332.——C. tenuiflorus, *Roxb. fl. Ind.* 1. *p.* 199; (*ed Wall.*) 1. *p.* 203 (exc. omnibus syn.——Peninsula Indiæ orientalis.

37. C. *Wallichii* (N. ab. E.:) umbella decomposita, radiis erectis apice trifidis brachiatis, spiculis spicatis disticho-alternis linearibus patentibus 14–20-floris compressiusculis, squamis late ovatis obtusis rufo-castaneis undique albo-marginatis muticis rigidis septemnervibus, nervis omnibus prominulis tribus dorsalibus viridibus, stylo trifido, caryopsi obovata trigona læviuscula, involucri triphylli foliolo uno umbella multo longiore, involucellis bracteolisque setaceis ciliatis, foliis linearibus margine culmique acute triquetri angulis scabris.—C. inundatus, *Herb. Roxb. et Ham.*; *Wall. cat. n.* 3342. *c.*——Ex Bengalia inferiori; *Wallich.*

Quanta quanta sit citata auctoritas, *Cyperus* tamen *inundatus* Roxburghii esse noster nequit ob stylum trifidum et caryopsin perfecte trigonam.

38. C. *marginellus* (N. ab E.:) umbella decomposita involucro tri- (4-) phyllo breviori, spicis compositis late pyramidalibus, spiculis lanceolatis 14–18-floris, squamis ovatis acutiusculis dorso nervosis pallide rubicundis margine angusto membranaceo albo cinctis, stylo trifido, caryopsi ovali trigona subtilissime punctulata, culmo acute triquetro lævi, foliis subradicalibus carinatis culmum subæquantibus.—β; spiculis imperfectis brevioribus patentibus, spicas subcylindricas struentibus.——China; *Vachell.* Ceylona; *Macrae.*

An *C. hexastachyii* varietas? Quod ad spiculas et spicularum distributionem simillimus *C. latifolio*, Poir.

39. C. *carnosus* (Heyne:) umbellæ decompositæ radiis brevibus densis simplicibus aut apice trifidis, spiculis spicatis alternatim confertis oblongo-lanceolatis compressiusculis 12–14-floris, squamis late ovatis obtusissimis muticis arcte imbricatis pallide ferrugineis margine omni albo membranaceo, dorso nervosis carina pallidiore, invollucellis squamiformibus, involucri triphylli foliolo altero longissimo culmum subcontinuante foliisque acute cul-

moque triquetris lævibus.—*Heyn. in Wall. cat. n.* 3355. *a* (an et reliqui eodem loco revensiti?)——Peninsula Indiæ orientalis.

Similis primo adspectu *C. canescenti* et formis quibusdam *C. proceri*, at differt evidentissime foliis culmum æquantibus triquetris spongiosis, facie plana, carina acuta, marginibusque lævibus.

§ 7. *Caryopsis trigona, stylo trifido. Spiculæ angustæ, basi ebracteatæ, in spicas simplices dispositæ.*—Alopecuroidei, *N. ab E.*

40. C. *canescens* (Vahl.:) umbella 6–8-radiata radiis brevibus densis, apice composite supradecompositeque spicatis, spica communi triangulari, propriis spiculisque divaricatis, his oblongis convexis sex-octofloris rigidis, squamis dense imbricatis ovatis mucronulatis nervosis glauco-fuscis, stylo trifido, caryopsi obovata trigona, involucro 6–7-phyllo umbella multo longiori foliisque glaucis margine scabris, culmo trigono angulis obtusis.—*Vahl. en.* 2. *p.* 355 (excl. syn. *Rheed.*); *R. et Sch. s. v.* 2. *p.* 211; *Wight. cat. n.* 1833.—Cyperus, *Wall. cat. n.* 3359. *b.*——In Oahu insula, in China, Peninsula Ind. or.

Habitu maxime accedit *C. polystachyo*, sed multo major, propiorque, quod ad inflorescentiam, *C. ligulari, ventricoso*, et confinibus.

41. C. *exaltatus* (Retz.) umbella decomposita, radiis universalibus apice corymboso-umbellatis e spicis pluribus pedicellatis erectis subcylindricis, spiculis imbricatis [vel undique patulis, *Arn.*] lanceolatis aut oblongo-lanceolatis compressis ebracteolatis 10–20-floris, squamis ovatis mucronatis lævibus rufo-castaneis nervo carinali viridi, duobus infimis minoribus sterilibus, stylo trifido, caryopsi ovali trigona utrinque acuta læviuscula alba, involucri penta-hexaphylli foliolis linearibus margine serrulato-scabris culmum triquetrum elatum subæquantibus, ochreis plerumque bifidis.—*Wight. cat. n.* 1834.—Cyperus, *Wall. cat. n.* 3343 (ex parte).—*α*; spicis brevioribus fastigiatis.—C. exaltatus, *Retz. obs.* 5. *p.* 11.—C. fastigiatus, *Rottb. gram. p.* 32. *t.* 7. *f.* 2; *Willd. sp. pl.* 1. *p.* 285; *R. et Sch. s. v.* 2. *p.* 220.—*β*; spicis præsertim mediis longioribus linearibus, spiculis brevioribus paucifloris (6–10-floris).—C. canaliculatus, *Retz. obs.* 5. *p.* 20?—*γ, bulbosus;* culmo humili radiisque umbellæ parvæ basi bulbosis nigro fuscis.——In Peninsula Indiæ or. locis humidis; in Bengalia inferiori; in Ceylona insula?

42. C. *altus* (N. ab E.:) umbella supradecomposita coarctata, partialibus corymboso-umbellatis propriisque trifidis, spicis laxis subdistiche spiculigeris, spiculis alternis bifariam patentibus lanceolatis compressis 20-floris ebracteolatis, squamis ovatis mucronatis latere lævibus rufis carina viridi 5–7-nervi, infima lanceolata acuminata bracteolam brevem referente, stylo trifido, caryopsi ovali trigona utrinque acuta alba, involucri 5–6-phylli foliolis 2–3 umbella longioribus angustis folioque radicali carinatis margine serrulato-scabris, culmo trigono lateribus canaliculatis, ochreis bifidis.—*Wight. cat. n.* 1835.—C. exaltatus, *Vahl. en.* 2. *p.* 366; *R. et Sch. s. v.* 2. *p.* 219 (excl. syn. *Rottb.*).—C. alopecuroides, *Roxb. fl. Ind.* 1. *p.* 208; (*ed Wall.*) 1. *p.* 211 (exc. syn.).——Peninsula Ind. or.

Proximus quidem *C. exaltato*, Retz., at differre tamen videtur notis allatis. [*C. alopecuroides*, Roxb., seu *C. acerosus*, Roxb. in cœt. merc. Ind. or. mus. tab. 731, et in Wall. L. n. 3343, *e*, vix hujus loci, sed ad *C. exaltatum*, Retz. pertinet.—ARN.)

43. C. *Roxburghii* (N. ab E.:) umbella decomposita supra decompositaque contracta radiis universalibus elongatis compresso-trigonis fasciculato-umbellatis e spicis pluribus fastigiatis cylindricis subsessilibus, spiculis imbricatis linearibns teretiusculis ebracteolatis 12–16-floris, squamis arcte imbricatis amplectentibus subrotundo-ovatis obtusis cum mucronulo latere membranaceis albidis dorso griseo-rufescentibus, stylo trifido stigmatibus elonga-

tis, caryopsi oblonga trigona utrinque acuta lævi, involucri pleiophylli foliolis binis exterioribus maximis foliisque radicalibus latis carina margineque serrulato-scabris, involucellis angustis umbellula brevioribus, culmo triquetro angulis superne spinuloso-scabris, lateribus concavis.—*Wight. cat. n.* 1836.—C. elatus, *Roxb. fl. Ind.* 1. *p.* 204; (*ed Wall.*) 1. *p.* 207 (excl. synon.).—C. maximus, *Roxb. in cœt. merc. Ind. or. mus. tab.* 728 (fide *Arnott*).——In Coromandelia et Benghala.

44. C. *digitatus* (Roxb.*:) umbella composita pluriradiata, radiis compressis strictis apice umbellatis e spicis 5–6 subæqualibus erectis teretibus a basi densissime imbricatis, spiculis lanceolatis compressis ebracteatis 4–14-floris, squamis ovatis obtusis mucronatis flavis dorso trinervibus nervo carinali viridi, stylo trifido, caryopsi ovali triquetra lævi pallida, involucellis diphyllis umbellula brevioribus, involucro subhexaphyllo, foliolis 4 umbella longioribus linearibus latiusculis foliisque margine rubris, culmo trigono lævi.—C. digitatus, *Roxb. fl. Ind.* 1. *p.* 205; (*ed. Wall.*) 1. *p.* 209; *N. ab E. in herb. Royle.*—Cyperus, *Wall. cat. n.* 3345.——Silhet.

C. digitatum, Roxburghii, esse puto, quoniam plurima congruunt, nomen ante omnia; idemque et formas majores occurrere docet, spiculis non modo 6-floris, sed 24-floris gaudentes.

45. C. *racemosus* (Retz.:) umbella decomposita fastigiata multiradiata, radiis semiteretibus strictis universalibus partialibusque apice umbellatis e spicis pluribus sessilibus linearibus strictis, spiculis imbricatis lineari-lanceolatis compressis subebracteolatis 5–10-floris, squamis ovatis mucronatis dorso fulvis rufove fuscis nervo carinali viridi latere flavis, stylo trifido, caryopsi ovali alba latere interiori concavo, involucro decaphyllo erecto foliolis e basi latiori attenuatis margine scabris 3–4 umbella longioribus, involucellis foliaceis umbellulam superantibus æquantibusve, foliis radicalibus lato-linearibus culmo triquetro firmo altoque longioribus.—*Retz. obs.* 4. *p.* 20; *Willd. sp. pl. p.* 287; *Vahl. en.* 2. *p.* 355; *R. et Sch. s. v.* 2. *p.* 211. *n.* 164; *Wight. cat. n.* 1837.—Cyperus, *Wall. cat. n.* 3341.—[C. digitatus, *Roxb. fl. Ind.* 1. *p.* 205; (*ed. Wall.*) 1. *p.* 209.—C. glomeratus, *Roxb. in cœt. merc. Ind. or. mus. tab.* 729.—Arn.]——Peninsula Indiæ orientalis. Penang.

[In nostris squamæ apice patulæ, cum mucrone: at in tab. picta Roxburghiana acuminatæ et subrecurvæ delineantur.—Huic valde quoad habitum et inflorescentiam affinis est *C. umbellatus*, Roxb. fl. Ind. 1. p. 205 (et in cœt. merc. Ind. or. mus. tab. 729); in quo umbellæ radii quidam breves umbellulas oligostachyas exinvolucratas gerentes, alii elongati apice umbelliferi involucello 3–4-phyllo: umbellula laxa, involucellum æquans, e radiis 3–4 composita brevibus apice spicas 3–4 lineares graciles subsessiles gerentibus; spiculæ parvæ 4–6-floræ arcte imbricantes et squamis late ovatis acutiusculis muticis dorso viridibus latere brunneis imbricatæ. Habitu omnino dissimilis *C. umbellato*, Burm. et Vahl., varietates quasdam deformatas magis quam ullam speciem typicam tamen referenti.—Arn.]

§ 8. *Caryopsis trigona stylo trifido. Spiculæ pauci-pluriflorœ capitatim aut spicatim confertæ basi bracteola suffultæ. squama infera in rhachi persistente, spiculis autem integris defluentibus. Rhachilla anguste marginata. Squamæ margine membraaceæ.*—Mariscoidei, *Lindl.*

a, *Capitati.*

46. C. *dubius* (Rottb.:) spiculis in capitulum ovatum basi compositum ebracteatum congestis oblongo-lanceolatis utrinque convexiusculis 6–8-floris, squamis imbricatis late ovatis obtusis herbaceo-chartaceis nervosis margine

* *C. digitatus* certe non huc, sed ad speciem sequentem referendus: *Cyperus* igitur *digitatus*, N. ab E., nuncupari potest *C. Roylei*.—Arn.

membranaceis nervoso-striatis pallidis, stylo trifido, caryopsi obtuse trigona, involucri subtetraphylli foliis tribus prælongis foliisque linearibus culmo sublongioribus lævibus, culmo trigono mediocri.—*Rottb. ic. p.* 20. *t.* 4. *f.* 5; *Roxb. fl. Ind.* 1. *p.* 188; (*ed. Wall.*) 1. *p.* 192; *Wight. cat. n.* 1838.—C. Kyllingoides, *Vahl. en.* 2. *p.* 312; *Willd. sp. pl.* 1. *p.* 273; *R. et Sch. s. v.* 2. *p.* 173 (excl. syn. *Pursh.*).—Cyperus monocephalus, *Roxb. in cæt. merc Ind. or. tab.* 707 (ex. *Arn*)—Cyperus, *Wall. cat. n.* 3326.—*Rumph. Amb.* 6. *t.* 2. *f.* 1.—— In Peninsula Ind. or., et Benghala.

Caryopsis trigona, punctulato-rugulosa, brunnea. Stylus bifidus altero ramo bifido, quo forsan deceptus bifidum perhibuit esse Rottbollius.

b, *Spicati, rhachi hirta.*

47. C. *obliquus* (N. ab E.:) umbella decomposita (supra decompositaque?) radiis compressis strictis apice trifidis radiolo medio basi composito, rhachi hirta, spiculis alternatim approximatis bifariam patulis linearibus brevibus teretiusculis bracteolo setacea plerumque brevioribus 5–7-floris, squamis arcte imbricatis late ovatis convexis apice rotundatis muticis aut infra apicem brevissime mucronulatis dorso refescentibus quinquenervibus latere lutescentibus margine membranaceo albo undique cinctis, stylo trifido?, involucellis nullis, involucri subpentaphylli foliolis duobus longissimis foliisque linearibus complicato-carinatis margine scabris nudis, culmo triquetro angulis scabris nudis.—Cyperus, *Wall. cat. n.* 3334. *c.*——In Kamoon.

Rhachibus hirtis et spiculis vix linea una longioribus distinctissima species, conferenda cum *C. piloso*, Vahl., cui spiculæ bilineares 12-floræ et spicæ horizontales brevissimæ in radiis longis tribuuntur; in nostra autem radii breves sunt et spicæ angustæ, media fere pollicari. *C. fimbriato*, cui jure proximum istum sensebis, differt: culmi angulis foliorumque margini haud ciliatis, spiculisque (eodem fere statu ætatis a me observatis) plus duplo minoribus, bracteolis longioribus præditis, pallidioribus.

48. C. *fimbriatus* (N. ab E.:) umbella decomposita involucro tri-tetraphyllo breviori, spicis sessilibus subfasciculatis, rhachi hirsuta, spiculis patulis lanceolatis squamisque fuscis obtusis, culmi triquetri angulis foliis (?) involucrisque ciliatis.—*Royle. herb. cyp. n.* 29.——Nepalia.

Ciliis culmum et involucra ornantibus distinctissimus omnium, *C. piloso*, Vahl., solo accedens, cut sola rhachis hirsuta.

49. C. *venustus* (R. Br.:) umbella supradecomposita decompositaque multiradiata, radiis universalibus partialibusque compressis apice umbellatim compositis e spicis pluribus sessilibus inferioribus radiatim patentissimis, spiculis approximatis confertisve patentissimis linearibus compressiusculis brevibracteolatis 12–15-floris, squamis ovatis obtusiusculis cum mucrone latere luteis carina trinervi virente, stylo trifido, caryopsi obovata trigona lævi pallida, rhachilla alata, auriculis apice demum solutis, involucri tri-octophylli foliolis alternis planis latis trinervibus foliisque margine serrulato-scabris, nonnullis umbella multo longioribus, culmo alto trigono lævi.—*Br. prod. fl. nov. Holl.* 1. *p.* 217; *Sieb. fl. nov. Holl. n.* 111; *Wight. cat. n.* 1839.—C. alopecuroides *Koen. apud R. Br.*—*Rheed. Mal.* 12. *t.* 42.——Nova Hollandia. Peninsula Ind. or. Java.

C. elatus, Linn., cujus exemplum Zeylanicum in Herb. cl. Lindleyi vidi, non differt nisi spicis spicularum densissimis, e spiculis in rhachi fere contiguis licet patentissimis, et rhachi crassiori multisulcata, quæ dubito an sufficiant ad distinguendam speciem. Quin, nostro *C. elati* nomen certo imposuissem, nisi cl. Brownii obstaret auctoritas, cui *C. elatum* cognitum fuisse, cum *venustum* suum describeret, vix dubium videtur. Præter *elatum* hunc, nimis, ut diximus, propinquum, *C. verticillatum*, Roxb., comparabis; differt hic autem spiculis paulo majoribus squamisque magis nervosis, quæ nostro non nisi in dorso trinerves, in latere autem altero luteæ, in altero pallide lutescentes.

50. C. *verticillatus* (Roxb.:) culmo folioso trigono, umbella decomposita et supradecomposita, spiculis spicatis confertis lineari-subulatis 8–9-floris brevibracteolatis spicisque partialibus infimis patentissimis, squamis ovatis obtusiusculis cum mucrone dorso 7-nervibus lateribus luteis, involucellis linearibus umbellas subæquantibus, involucri polyphylli foliolis plerisque umbellam superantibus foliisque lato-linearibus margine serrulato-scabris.—*Roxb. fl. Ind.* 1. *p.* 206; (*ed. Wall.*) 1. *p.* 209 (excl. syn.); *N. ab E. in herb. Royle. cyp. n.* 12.——In Peninsula Ind. or.; *Roxburgh.* Nepala; *Royle.*

Squamæ spicularum pro spiculæ magnitudine sat magnæ, alternæ, arcte imbricatæ, apicibus erecto-patulis. Spicula 3½ 4 lineas longa.

[In tabula ejus stirpis picta Roxburghiana (in cœt. merc. Ind. or. mus. t. 1319) radii universales apice spicas 5–8 cylindraceas imbricatas unciam longas verticillatas patentes subsessiles gerunt: e centro umbellulæ seu verticilli oritur aut spica unica erecta subsessilis cæteris similis, aut radii 1–2 spicas subæquantes apice oligo- (2–3-) stachyi; spiculæ adpressæ lanceolatæ 8–12-floræ lineam longæ. Inflorescentia itaque diversissima differt a *C. venusto*, nec plantam supra a Neesio descriptam esse Roxburghianam credere possum. —ARN.]

§ 9. *Caryopsis trigona, stylo trifido. Squamæ spicularum obtusæ, caryopsim æquantes. Spiculæ lineares, laxæ, capitatim aut digitatim dispositæ, umbella composita decompositave.*—Irioidei, *Lindl.*

51. C. *parviflorus* (Vahl:) umbella decomposita, radiis universalibus apice cymosis aut abbreviatis fastigiatis, spiculis spicatis alternis ovalibus obtusis tri-quinquefloris, squamis obovatis infra apicem obtusissimum retusumve mucronatis caryopsim æquantibus, stylo trifido, caryopsi triquetra, involucri subpentaphylli foliolis ternis foliisque linearibus elongatis margine serrulato-scabris, culmo trequetro.—α; umbella triquetro.—β; umbella propter radios abbreviatos contracta, quasi e spicularum fasciculo constante.—*Wight. cat. n.* 1840.—C. parviflorus, *Vahl. en.* 2. *p.* 352; *R. et Sch. s. v.* 2. *p.* 210.—C. umbellatus, *Roxb. fl. Ind.* 1. *p.* 205; (*ed. Wall.*) 1. *p.* 208 (nec *Burm.*); *N. ab E. in cyp. Royl. n.* 12.—Cyperus, *Wall. cat. n.* 3360. *c* (fide *Arn.*).——Peninsula Ind. or.; *Wight.* Nepalia; *Royle.*

Similis *C. Iriæ*, præsertim quod ad squamas attinet obtusas, infra apicem mucronatas oblique 5-nerves latere flavicantes; sed differt spiculis parvis alternis distantibus in spicas angustas fasciculatim subumbellatas dispositis.

[*C. umbellatus*, Roxb. (in cœt. merc. Ind. o. mus. tab. 729) spiculæ squamarum forma et longitudine, aliisque multis notis valde differt. Plantam a Royleo lectam ignoro, at Wightiana *C. Iriæ* nimis affinis est, et ex eadem radice spiculas tam plurifloras quam paucifloras profert.—ARN.]

53. C. *Iria* (Linn.:) umbella supradecomposita, radiis spicatim compositis, spiculis imbricatis oblongo-linearibus pallidis plurifloris, squamis obovatis obtusis submucronulatis oblique 3–5-nervibus caryopsin æquantibus, caryopsi triquetra mucronulata, culmo triquetro, foliis linearibus lævibus caulem subæquantibus.—*Linn.; Willd. sp. pl.* 1. 1. *p.* 287; *Link. hort. Ber.* 1. *p.* 308; *Roxb. fl. Ind.* 1. *p.* 201; (*ed. Wall.*) 1. *p.* 204; *Presl. rel. Haenk.* 3. *p.* 174; *Wight. cat. n.* 1841.—Cyperus, *Wall. cat. n.* 3360.—C. Santonici, *Rottb. gram. p.* 41. *t.* 9. *f.* 1.—*Rheed. Mal.* 12. *t.* 56; *Pluk. t.* 191. *f.* 7.—β; umbella rigidula contracta.—C, chrysomelinus, *Link. h. Ber.* 1. *p.* 305.—γ, *subtilis;* culmo 4–6 poll. longo gracili, foliis involucrisque filiformibus, umbella parva contracta ovali aut obovata.—*Wight. cat. n.* 1841. *b.*—Cyperus, *Wall. cat n.* 3361.——α; Manila insula. China, ad Macao; *Vachell.* Peninsula Ind. or.—β; in hortis colitur, semine e Nepalia adlato.—γ; in Mysore; *Wight.*

53. C. *difformis* (Linn.:) umbella subdecomposita breviradiata spiculis globoso-capitatis confertissimis lanceolato-linearibus compressis parvis 8–14-floris, squamis subrotundis obtusis margine late membranaceis albis carina viridi fuscave trinervi, stylo bifido, caryopsi ovali trigona punctulato-aspera squamam æquante, involucro 2–3-phyllo, foliolo altero binisve longissimis, foliis lineari-acuminatis culmum trigonum subæquantibus.—*Linn. sp. pl. p.* 74; *Willd. sp. pl.* 1. *p.* 280; *R. et Sch. s. v.* 2. *p.* 193; *mant.* 2. *p.* 110; *Link. h. Ber. p.* 304; *Presl. in rel. Haenk.* 3. *p.* 169; *Rottb. gram. p.* 24. *t.* 9. *f.* 2.; *Roxb. fl. Ind.* 1. *p.* 195; (*ed Wall.*) 1. *p.* 199; *in cœt. merc. Ind. or. mus. tab.* 716 (fide *Arn.*); *Wight. cat. n.* 1824.—C. protractus, *Del.*; *Link. l. c. p.* 305.—Cyperus, *Wall. cat. n.* 3363.——Per omnem Indiam, Novam Hollandiam, Insulam St Mauritii, Ægyptum usque in Græciam et Lombardiam; *Presl.* In Manilla; *Haenke; Meyen.* In China; *Vachell.*

54. C. *distans* (Linn.:) umbella decomposita, radiis universalibus apice umbellatis e spicis 3–4 sessilibus patentissimis, spiculis alternis patentissimis linearibus subduodecim-floris bracteola setacea suffultis squamis alternis distantibus late ovalibus obtusissimis margine membranaceis caryopsin oblongam obtusam trigonam æquantibus, stylo trifido, involucri subpentaphylli foliolis binis ternisve umbella longioribus foliisque planis linearibus latiusculis margine nervisque scabris, culmo (1–3-pedali) triquetro.—*Linn. suppl. p.* 103; *Jacq. ic. pl. rar.* 2. *t.* 299; *coll.* 3. *p.* 189; *Willd. sp. pl.* 1. *p.* 288; *Vahl. en.* 2. *p.* 363; *R. et Sch. s. v.* 2. *p.* 216. *n.* 176; *Pal. de Beauv. flore d'Ow.* 1. *p.* 35. *t.* 20; *Roxb. fl. Ind.* 1. *p.* 207; (*ed. Wall.*) 1. *p.* 210; *Wight. cat. n.* 1343.—C. elatus, *Rottb. gram. p.* 37. *t.* 10.—C. laxus. *Vahl. en.* 2. *p.* 362; *R. et Sch. s. v.* 2. *p.* 217.—Cyperus, *Wall. cat. n.* 1843.——Peninsula Indiæ orientalis locis udis, Malabar. Insula S. Crucis. In regno Oware et Benin; *Palis. de Beauv.*

II. PAPYRUS. *Link.*

1. P. *elatus* (N. ab E.:) culmo folioso prælto trigono, umbella supradecompositaque multiradiato, spicis partialibus pluribus sessilibus teretibus inferioribus brevioribus patentissimis, rachi crassa sulcata, spiculis confertissimis patentissimis linearibus compressis brevibracteolatis 12–floris, squamis ovatis obtusis cum mucrone latere lutescentibus dorso purpurascentibus carina virente, rachillæ auriculis apice liberis demum solubilibus, involucri polyphylli foliolis alternis planis latis trinervibus foliisque margine serrulato-scabris, nonnullis umbella multo longioribus.—C. elatus, *Linn. am. acad.* 4. *p.* 301; *R. et Sch. s. v.* 2. *p.* 217.——Ceylona insula.

Species distincta spicis partialibus inferis ad basin centralis quasi verticillatis dense congestis, patentissimis, qua nota ad *Cyperum verticillatum* nimis approbinquatur. Rachillæ auriculæ latæ, caryopsin amplectentes et, cum ea perficitur, a rhachilla solutæ *Pypyri* generis characterem præ se ferunt.

2. P. *Pangorei* (N. ab E.:) culmo aphyllo trigono, umbella decomposita, spiculis suflabellatis alternis patulis linearibus 16–40-floris, squamis ovali-oblongis obtusis carina trinervi margine emembranaceis, involucellis bractealibus alternis ovato-attenuatis ochreolas æquantibus, involucro triphyllo umbellam æquante, squamulis internis persistentibus.—*Wight. cat. n.* 1844.—Cyperus Pangorei, *Rottb. gram. p.* 31. *t.* 7. *f.* 3; *Vahl. en.* 2. *p.* 345; *R. et Sch. s. v.* 2. *p.* 203; *Roxb. fl. Ind.* 1. *p.* 202; (*ed. Wall.*) 1. *p.* 206 (excl. syn. *Willd.*)—[C. tegetum, *Roxb. fl. Ind.* 1. *p.* 208; (*ed. Wall.*) 1. *p.* 211; *in cœt. merc. Ind. or. tab.* 1022.—ARN.]—Cyperus, *Wall. cat. n.* 3351, 3352, *a* (ex parte).——Peninsula Ind. or.

Inter species culmo aphyllo instructas hæc nostra differt spiculis langioribus magis patulis, squamis haud ita rigidis, involucro stricto latiusculo umbellam non superante.

[Exemplis perplurimis eductus sum involucri foliola nunc umbella dimidio breviora nunc eam duplo superata esse, characteresque minimi valoris præbere. Spiculæ quamvis sæpissime unciam longæ, quandoquidem 2–3 lineas vix excedunt. *Cyp. Pangorei*, Roxb., vix hujus loci.—Species valde affinis, a Neesio omissa, est. *P. tegetiformis*, Arn.: culmo aphyllo trigono, umbella decomposita, spiculis alternis attenuatis approximatis patulis linearibus 20–30-floris, squamis ovali-oblongis obtusis læte brunneis, involucellis bractealibus alternis, involucro 3–4-phylli umbella 4–5-plo breviores foliolis lanceolatis squamæformibus.—Cyperus tegetiformis, *Roxb. hort. Bengh.; in cœt. merc. Ind. or. mus. tab.* 1321; *Wall. cat. n.* 3351, *a* (ex parte).—C. nudus, *Roxb. fl. Ind.* 1. *p.* 209; (*ed. Wall.*) 1. *p.* 213.—Arn.]

3. P. *dehiscens* (N. ab E.:) culmo basi unifolio superne triquetro, umbella decomposita radiis gracilibus bi-trifidis, spiculis 5–6-nisve approximatis alternis patentibus linearibus 12–16-floris, squamis oblongis obtusis patulis dorso viridi-nervosis margine rufescentibus, involucellis bractealibus exiguis lanceolatis squamiformibus, involucro triphyllo umbellam superante, squamulis interioribus solubilibus.—Cyperus corymbosus, *herb. Heyn. in Wall. cat. n.* 3352 (ex parte, cum P. Pangorei *et* corymboso).——Peninsula Ind. or.; *Heyne*.

Species, culmo folioso, spiculis in umbella laxa humili patulis, involucri omnibus foliolis umbellam duplo et amplius superantibus, squamis alternis patulis, invollucellis fere nonnisi squamulis, quales et infra spiculas occurrunt, satis distincta.

[Huc pertinere videtur *Cyp. Pangorei*, Roxb. (a Neesio male ad *Pap. Pangorei* relatum) seu *C. gangeticus*,, Roxb. tab. pict. n. 1111, igiturque forsan *Cyperus*, Wall. L. n. 3351, *a* (ex parte).—Arn.]

4. P. *corymbosus* (N. ab E.:) culmo aphyllo trigono, umbella supradecomposita radiis gracilibus, spiculis in apice radiolorum fasciculatis confertis lineari-subulatis 12–14-floris, squamis ovalibus obtusis convexis dorso viridi-nervosis margine pallidis disco rufescentibus, involucellis bractealibus confertis e basi scariosa ovata setaceis ochreolas superantibus, involucro tri-tetraphyllo stricto umbellam æquante, squamulis interioribus liberis deciduis.—Cyperus corymbosus, *Rottb. gram. p.* 42. *t.* 7. *f.* 4; *Willd. sp. pl.* 1. *p.* 287 (excl. syn. *Thunb.*); *herb. Heyn. in Wall. cat. n.* 3352. *a* (ex parte).—C. tegetum, *Roxb. fl. Ind.* 1. *p.* 208; *ed Car. et Wall.* 1. *p.* 211.—C. monophyllus, *Vahl. en.* 2. 342; *R. et Sch. s. v.* 2. *p.* 209.——Peninsula Ind. or.

Inflorescentia gracili, involucellis ob radiolos fasciculatos confertis setaceo-acutatis, spiculis circiter 4-linearibus angustis in fasciculos densos ad apices radiolorum collectis involucro umbellam æquante facile distinguitur. *C. corymbosus*, Vahl. ad exempla Capensia descriptus, aliena est species.

[Quamvis auctor huc absque dubio tradit *C. tegetum*, Roxb. (in cœt. merc. Ind. or. mus. tab. 1022, et forsan Wall. cat. n. 3332. *a.*) descrimina haud levia sese ostendunt: in planta Roxburghiana, radioli minime approximati at revere alterni et paullo distantes; involucella itaque non conferta ut in *P. corymboso*: nec dubito quin eadem sit species ac *P. Pangorei*.—Arn.]

III. MARISCUS. *Link.*

Differt a *Cypero* spiculis a squamis inferioribus articulo solubilibus, rachi residua post lapsum spicularum quasi paleacea remanente.

1. M. *paniceus* (Link.?:) culmo trigono gracili, spicis ternis quinisve oblongis sessilibus pedunculatisque, spiculis primum subulatis falcato-recurvis dein ovato-oblongis trigonis erecto-patulis apice subulato recurvo, squamulis duabus spicularum infimis æqualibus ovatis, inferiori acuta, involucro di-tetraphyllo umbella longiore.—*Link. hort. Ber.* 1. *p.* 323?; *Wight. cat. n.* 1845. —M. biglumis, *Gærtn. de fr. et sem.* 1. *p.* 12. *t.* 2. *f.* 8?—Mariscus, *Wall. cat. n.* 3434.—Kyllingia panicea, *Linn. suppl. p.* 105; *Rottb. gram. p.* 15. *t.* 4. *f.* 1; *Willd. sp. pl.* 1. *p.* 257.——Peninsula Ind. or., Ceylona.

Planta Vahlii hujusce nominis, cui "bracteæ setaceæ spiculis breviores," diversa est species. Probabiliter autem ad hanc referenda est var. illa: culmo foliisque filiformibus spicis tantum 2–4, a Vahlio ad *Mariscum umbellatum* laudata.

2. M. *pictus* (N. ab E.:) culmo trigono, umbella simplici pluriradiata, spicis densis subsessilibus, spiculis erectis imbricatis lanceolatis bifloris, squamula spicularum involucrali (sterili infera) brevissima acuta, fertilium inferiori ovali dorso quinquenervi latere albida purpureo-irrorata superiori duplo fere breviore.—*Wight. cat. n.* 1846.—Mariscus, *Wall. cat. n.* 3435, *a* ?—Kyllingia bulbosa, *Koen. apud Vahl. en.* 2. *p.* 376 (*sub* M. umbellato?)——In Benghala. In Peninsula Ind. or.

Differt a *Marisco cyperino* β**, spiculis majoribus purpuro irroratis, squama fertili inferiori duplo fere breviore quam superiori, ovali, obtusissima, e tenuiori membrana constructa, dorso nervis quinis prominulis magis distantibus divisa. *Kyllingia cyperina*, Retz., tam ad hunc quam ad *M. cyperinum* nostrum pertinere potest.

3. M. *cyperinus* (Vahl.:) culmo trigono, umbella simplici pluriradiata, spicis cylindricis sessilibus pedunculatisque, spiculis erectis uni-bifloris subulatis, squamula spicularum involucrali brevi lanceolata, fertilium inferiori dorso trinervi latere immaculata.—α; spiculis unifloris.—*Wight. cat. n.* 1847 (*a*, umbella radiata, et *b*, umbellæ spicis sessilibus).—Mariscus, *Wall. cat. n.* 3437.—β; spiculis bifloris.—β*; spicis pedunculatis.—β**; spicis omnibus subsessilibus.—M. cyperinus, *Vahl. en.* 2. *p.* 377; *R. et Sch. s. v.* 2. *p.* 298.—M. Sieberianus, *N. ab E. in Sieb. herb. Maur.* ii. *n.* 3 (ex parte).—Kyllingia cyperina, *Retz. obs.* 6. *p.* 21; *Willd. sp. pl.* 1. 1. *p.* 258 (umbellæ spicis sessilibus).—K. umbellata, *Roxb. fl. Ind.* 1. *p.* 182; (*ed. Wall.*) 1. *p.* 186 (excl. syn. *Linn.*, *Rottb.*, et *Vahl.*)—*Rheed. Mal.* 12. *t.* 63.——α; In Peninsulæ alibique Indiæ humidis. China, Ins. Ascens.—β; In China; *Vachell.* In Ins. St Maur.; *Sieber.*

Species, a *M. umbellato*, Vahl., satis distincta spiculis longioribus erecto-patulis, nec reflexo-imbricatis. Cæterum habitu satis conveniunt hæc atque illa species.

4. M. *umbellatus* (Vahl.:) culmo trigono, umbella simplici pluriradiata, spicis oblongo-cylindraceis, plerisque pedunculatis densissimis, spiculis retrorsum declinato-imbricatis unifloris lanceolato-subulatis, squamula involucrali lanceolata, inferiorum lineari-elongata, fertili nervoso-striata, quarta rudimentali.—*Vahl. en.* 2. *p.* 376; *R. et Sch. s. v.* 2. *p.* 247 (excl. syn. *Pursh.* et aliis); *Link. h. Ber.* 1. *p.* 324.—M. Sieberianus, *N. ab E. in Sieb. Maur.* ii. *n.* 2 (ex parte).—Scirpus cyperoides, *Linn. mant. p.* 181.—Kyllingia umbellata, *Linn. suppl. p.* 105; *Rottb. gram. p.* 15. *t.* 4. *f.* 2; *Willd. sp. pl.* 1. 1. *p.* 257.—R. sumatrensis, *Retz. obs.* 4. *p.* 13.——In Peninsula aliisque locis Indiæ orient. In insula S. Mauritii; *Sieber.*

Spiculis parvis, leni arcu recurvis retrorsumque imbricatis, valvula fertili pulchre striata (11-nervi), spicis ratione plantæ parvulis, involucro longissimo facile distinguitur.

5. M. *dilutus* (N. ab E.:) culmo teretiusculo obtuse trigono, umbella composita multiradiata, spicis globosis, spiculis patentissimis linearibus 6–8-floris, squamula involucrali brevi setacea, involucri subpentaphylli foliolis plerisque elongatis involucellisque margine denticulatis, vaginis basi reticulato-lacunosis.—*Wight. cat. n.* 1848.—Mariscus, *Wall. cat. n.* 3439.—Cyperus dilutus, *Vahl. en.* 2. *p.* 357; *R. et Sch. s. v.* 2. *p.* 214.—C. compactus, *Retz. obs.* 5. *p.* 10; *Willd. sp. pl.* 1. *p.* 271; *R. et Sch. s. v.* 2 *p.* 165.—C. congestus, *Poir. enc. meth. suppl.* 7. *p.* 239.—[C. spinulosus. *Roxb. fl. Ind.* 1. *p.* 203; (*ed. Wall.*) 1. *p.* 207; *in cœt. merc. I. o. mus. tab.* 707.—ARN.]——In Peninsula Ind. orient. ulteriori, Rangoon, Penang, Singapur.

IV. KYLLINGIA. *Linn.*

1. K. *monocephala* (Linn.:) capitulo subsimplici globoso dense compacto, spiculis unifloris diandris dimidiato-ovatis acuminatis striatis membranaceo-carinatis carina ciliatis albidis, involucro triphyllo capitulo multo longiore, caryopsi obovata, foliis angustis linearibus margine scabris.—*Linn. suppl. p.* 104; *Vahl. en.* 2. *p.* 379; *R. et Sch. s. v.* 2. *p.* 236; *Roxb. fl. Ind.* 1. *p.* 180; (*ed. Wall.*) 1. *p.* 185; *Link. h. Ber.* 1. *p.* 325; *Presl. in rel. Hænk.* 3. *p.* 182; *Rottb. gram. p.* 13. *t.* 4. *f.* 4; *N. ab E. in cyp. Royl. n.* 35. *et* 37; *Wight. cat. n.* 1849.—Kyllingia, *Wall. cat. n.* 3443.—*Rheed. Mal.* 12. *t.* 53; *Rumph. Amb.* 6. *t.* 8. *f.* 2.——In Peninsula aliisque locis Indiæ orientalis, frequens. In China; *Meyen; Vachell.* In Nepalia; *Royle.*

2. K. *brevifolia* (Rottb.:) capitulo simplici globoso sessili dense compacto, spiculis unifloris triandris dimidiato-ovatis acuminatis striatis carina viridi-trinervibus serrulato-scabris margine griseis, involucro triphyllo capitulo sex-octuplo longiore, caryopsi obovata, foliis angustis linearibus culmo brevioribus.—*Rottb. gram. p.* 13. *t.* 4. *f.* 3; *Willd. sp. pl.* 1. *p.* 256; *Vahl. en.* 1. *p.* 380; *R. et Sch. s. v.* 2. *p.* 237; *N. ab E. in herb. Royl. n.* 38; *Wight. cat. n.* 1850. —Kyllingia, *Wall. cat. n.* 3440.—β; foliis latioribus (2 lin. latis), carina squamarum uninervi glabriuscula.—*Wight. cat. n.* 1850, *b.*——Peninsula Ind. orient.

Est species, *K. monocephalæ* sane simillima, sed distincta carina valvularum viridi trinervi plerumque spinuloso-serrata lateribusque haud albis sed ex albo subrufis, foliis brevioribus. *K. cruciformis*, Schrad., differt spiculis monandris.

3. K. *melanosperma* (N. ab E.:) capitulo simplici ovato-subrotundo sessili dense compacto, spiculis unifloris triandris, squamis ovato-lanceolatis acuminatis septemnervibus nervis ternis lateralibus costiformibus, carina spinuloso-serratis, caryopsi oblonga latere subcostata punctata atra, involucro tri-tetraphyllo capitulo multo longiore margine scabro, foliis brevissimis, culmo alato-triquetro.—*Wight. cat. n.* 1851.——Peninsula Indiæ orientalis; *Rottler.*

Differt a *K. brevifolia* statura altiore, culmo fortiori acutissime triquetro, foliis involucralibus 4–5-pollicaribus, caulino vix pollicari, squamis spiculæ in utroque latere tricostatis. A *K. cruciformi*, cui propinquior, differt culmo argutissime triquetro, foliis multo brevioribus, squamis latere tri-nec bicostatis. Ab utraque omnibusque confinibus insuper caryopsi duplo fere longiore quam lata, compressa, utroque latere obsolete costulata, aterrima, punctulis exasperata.

* K. *cruciformis* (Schrad.:) capitulo globoso sessili solitario, involucro capitulo multo longiori triphyllo, spiculis monandris, squamis ovatis acuminato-mucronatis apice recurvo quinquenervibus carina spiculoso—scabris, caryopsi obovata lutea, culmo gracili foliisque culmo brevioribus angustis lævibus.—*Schrad. in Schult. mant.* 2. *p.* 137. *n.* 1, *a* (quoad exempla Americana).—K. uncinata, *Link. h. Ber.* 1. *p.* 325 (excl. syn. K. brevifoliæ, *Rottb.* et *Auct.*)—Cyperus Kyllingæoides, *Sieb. fl. Martin. n.* 16.——Bahiæ humidis; insulæ St Vinc.; *Macrae.*

4. K. *cylindrica* (N. ab E.:) capitulis subternis densis, medio cylindrico obtuso, involucro 3–4-phyllo, binis foliolis capitulo multo longioribus, spiculis monandris, squamis ovatis acutis rectis pallide viridibus purpura irroratis, carina lævi, foliis planis culmum subæquantibus involucrisque margine et carina scaberrimis.—*N. ab E. in herb. Royle. n.* 39.—β, monostachya.—Kyllingia, *Wall. cat. n.* 3442.——Nepal.

Sola, quæ isti comparanda, *K. odorata*, Vahl., gaudet staminibus duobus, spiculisque minoribus carina ciliolatis pallidioribus.

5. K. *triceps* (Linn.:) capitulis ovato-subrotundis subternis sessilibus me-

dio majori, spiculis diandris unifloris, squamis ovato-oblongis acutis membranaceo-albis lævibus carina subciliatis quinquenervibus nervis tenuibus, caryopsi oblongo-ovali compressa pallida, involucro tri-tetraphyllo foliolis ternis capitulo multo longioribus foliisque linearibus culmum trigonum superantibus margine scabris.—α; culmo altiori (sæpe gracili).—*Wight. cat. n.* 1852.—K. triceps, *Rottb. gram. p.* 14. *t.* 4. *f.* 6; *Willd. sp. pl.* 1. 1. *p.* 256; *R. et. Sch. s. v.* 2. *p.* 238 (cum syn. plerisque); *Roxb. fl. Ind.* 1. *p.* 181; (*ed. Wall.*) 1. *p.* 186.—Kyllingia, *Wall. cat. n.* 3441.—β; pumila.—*N. ab E. in cyp. Royle. n.* 36; *Wight. cat. n.* 1852, *b.*——α et β; in Peninsula Indiæ orientalis.—β; in Nepalia.

V. COURTOISIA. *N. ab E.*

1. C. *cyperoides* (N. ab E.)—*N. ab E. in Ed. n. ph. journ.* (1834) *n.* 34. *p.* 262; *Wight. cat. n.* 1853.—Kyllingia cyperoides, *Roxb. fl. Ind.* 1. *p.* 182; (*ed. Wall.*) 1. *p.* 187. *n.* 4.——In Peninsula Ind. or.

VI. REMIREA. *Aubl.*

1. R. *Wightiana* (Wall.:) pedunculo communi capitulum pyramidale basi compositum æquante, foliis undique patulis.—*Wall. cat. n.* 3378; *Wight. cat. n.* 1854.——Peninsula Ind. or.; *Wight.*

Persimilis quidem *Remireæ maritimæ*, sed differt, præter characteres supra expositos, spiculis paulo majoribus, squama penultima apice convoluto-acuminata distinctius nervosa.

TRIB. II. HYPOLYTREÆ. *N. ab E.*

VII. ANOSPORUM. *N. ab E.*

1. A. *monocephalum* (N. ab E.)—*N. ab E. in Ed. n. phil. journ.* (1834), *n.* 34. *p.* 263; *Wight. cat. n.* 1855.—Cyperus, *Wall. cat. n.* 3441.—C. monocephalus, *Roxb. fl. Ind.* 1. *p.* 188; (*ed Wall.*) 1. *p.* 193.—C. monocephaloides, *Roxb. in cœt. merc. Ind. or. mus. tab.* 1318 (fide *Arn.*)——Benghalia. Silhet. Peninsula Ind. or.

VIII. HEMICARPHA. *N. ab E.*

1. H. *Isolepis* (N. ab E.)—*N. ab E. in Ed. n. phil. journ.* (1834), *n.* 34. *p.* 263; *Wight. cat. n.* 1856.—*Roth.? nov. pl. sp. p.* 29.——Peninsula Indiæ orient; *Wight.*

IX. LIPOCARPHA. *R. Br.*—Hypoelytrum. *Lestib.*

1. L. *lævigata* (N. ab E.:) capitulo 5–7-stachyo, squamis communibus cuneiformi-unguicularibus subtrilobis cuspidulatis, spicula propria bivalvi, valvulis lanceolatis æqualibus convolutis, involucro diphyllo capitulo longiore, culmo compresso-trigono trigonove basi folioso.—Tunga lævigata, *Roxb. fl. Ind.* 1. *p.* 183; (*ed. Wall.*) 1. *p.* 188; *Schult. mant.* 2. *p.* 134.—Schœnus lævigatus, *Roxb. in cœt. merc. I. o. mus. tab.* 702 (fide *Arn.*)—α; culmo compresso trigono altiori, spiculis 6–7-nis, squamis uninervibus aut obsolete trinervibus.—(*, lacinia media producta.—Lipocarpha, *Wall. cat. n.* 3445 (excl. *a.*)—**, squamis leniter trilobis.—Lipocarpha lævigata, *N. ab E. in Cyp. Royl. n.* 55.)—β; culmo trigono humiliore, spiculis 4–5, squamis distincte trinervibus, lobo medio longe producta.—*Wight. cat. n.* 1857.—Lipocarpha, *Wall. cat. n.* 3445, *a.*——In vallibus humidis Penins. Ind. or. In Ceylona.

2. L. *triceps* (N. ab E.:) capitulo tristachyo, squamis communibus obovato-spathulatis trilobis venoso-nervosis brunneo-fuscis, spicula propria bivalvi, valvulis oblongis convolutis exteriori quinquenervi, involucro diphyllo

capitulo multo longiore, culmo compresso-trigono.—*Wight. cat. n.* 1858.—α; foliis lineari-elongatis.—Lipocarpha, *Wall. cat. n.* 3444. *a.*—Tunga triceps, *Roxb. fl. Ind.* 1. *p.* 183; (*ed. Wall.*) 1. *p.* 187.—Hypolyptum sphacelatum *et* filiforme, *herb. Heyn. in Wall. cat. n.* 3444. *c.* (nec *Vahl*).—β, brevifolia; foliis vaginis multo brevioribus, vaginarum inferiorum brevissimis.—Hypœlyptum Zeylanicum, *Heyn. in Wall. cat. n.* 3402. *b.*——Coromandelia.

[Nescio quomodo differt *Hyp. sphacelatum*, Vahl.: Rottlerus hoc nomen exemplis Heyneanis imposuit, etiamque exempla ad Vahlium misit; nec descriptio a Rottlero data multum differe videtur. Patria ipsissima.—ARN.]

X. HYPOLYTRUM. *Richard.*—Beera, *P. de Beauv.*—Albickia, *Presl.*

1. H. *giganteum* (Wall.:) capitulis corymbosis, spiculis ovatis, caryopsi patente ovato-subrotunda cum styli basi conica acuta confluente obsolete lacunulosa, perigynii bivalvis valvulis cymbiformibus.—*Wall. cat. n.* 3404; *Wight. cat. n.* 1859.—H. latifolium, *Rich.* (ex *Arn.*)—Tunga diandra, *Roxb. fl. Ind.* 1. *p.* 184; (*ed. Wall.*) 1. *p.* 188.—Schœnus nemorum, *Vahl.* (ex *Arn.*)—*Rheed. Mal.* 12. *t.* 58 (ex *Arn.*)—In Martabania, Chappedong oræ Tenassarim, Silhet; *Wallich.* E. Ceylona et Indiæ Peninsula; *Wight.* Amboina; *Roxburgh.*

Ab *H. schœnoide*, sive *Albickia schœnoide*, Presl, vix notis sufficientibus differt. Inflorescentia gaudet magis corymbosa, fructu breviori patentissimo nitidulo pallido obiter scrobiculato utrinque acutiusculo, rostro parum distincto.

XI. FUIRENA. *Linn.*

Duæ sunt subdivisiones; secundæ species in India orientali non occurrebant.

§ 1. *Perigynii foliola interiora unguiculata.*—Vaginaria.

Fuirenæ Vaginariæ flos e tribus verticillis constat; scil. 1*mo*, Calyce trisepalo, sepalis setaceis. 2*do*, Corolla tripetala, petalis unguiculatis cum sepalis alternis, 3*tio*, Staminibus tribus sepalis oppositis angulisque ovarii respondentibus. Igitur, 4*to*, Carpellis tribus petalis oppositis planis valvatim connatis, apicibus stigmaticis totidem liberis cum plano medio utriusque lateris congruentibus.

1. F. *ciliaris* (Roxb.:) spiculis capitatis axillaribus terminalibusque hirsutis, capitulis pedunculatis simplicibus compositisve, terminali geminato, squamis carinatis cristatis arista lamina breviori lateribus bi-tricostatis apicem versus membranaceis, perigynii foliolis interioribus membranaceis unguiculatis obovatis subretusis aristulatis ciliatis trinervibus, culmo oblique tetragono, vaginis foliisque hirsutis.—*Roxb. fl. Ind.* 1. *p.* 180; (*ed. Wall.*) 1. *p.* 184; *Wight. cat. n.* 1860.— Fuirena, *Wall. cat. n.* 3544.—Scirpus ciliaris, *Linn. mant. p.* 182; *Willd.? sp. pl.* 1. *p.* 309; *Herb. Klein. et Rottl.*——In locis humidis litoris Coromandel, inter Scirpos; *Roxburgh.* Aliis in locis Peninsulæ; *Wight.* In Silhet et in Promæ montibus; *Wallich.* In Ceylonæ insulæ locis humidis; *Macrae.*

Utrum hæc, an potius *F. Rottboellii* quam diximus, *Scirpus ciliaris*, Linn., existimanda sit, e descriptione Linneana vix dispiciandum. Moverunt me autem, hanc ut isto nomine appellarem, verba Linneana: "arista squamis breviore" et signum dubitationis, a Rottbellio synonymo Linneano adjectum.

2. F. *pentagona* (W. & A.:) spiculis capitatis axillaribus terminalibusque pubescenti-hirtis obscuris, capitulis pedunculatis compositis, terminali gemino, squamis planis membranaceis aristatis arista lamina dorso trinervi breviori, perigynii foliolis interioribus membranaceis unguiculatis ovalibus aristulatis glabris trinervibus, culmo gracili quinquangulari, vaginis glabris, supremis ore foliisque superioribus supra hirsutulis.—*Wight. cat. n.* 1861.—Fuirena, *Wall. cat. n.* 3542.—F. glomerata, *R. Br. prodr. fl. Nov. Holl.* 1. *p.* 220. ——Peninsula Indiæ orientalis; *Wight.* Ceylona; *Macrae.*

Magis accedit ad *F. Rottboellii* quam ad *ciliarem;* differt a *F. Rottboellii* squamis spiculæ majoribus arista breviori præditis, culmo altiore, capitulis pedunculatis. A *F. ciliari* distinguitur: squamis fusco-olivaceis nec ferrugineis, haud carinatis, dorso trinervibus, nec costis 2–3-nis in utroque latere instructis, totis ferme membranaceis neque ad latus tantummodo superius, setula gracili abrupte aristulatis neque in acumen subulatum recurvum exeuntibus, foliis vaginisque interioribus glabris nec omnibus hirsutis, culmo gracili altiori (1–1½-pedali) acute quinquangulari.

3. F. *Rottboellii* (N. ab E.:) spiculis capitatis axillaribus terminalibusque hirsutis, capitulis subsessilibus, terminali composito ovali, squamis membranaceis planis trinervibus aristatis arista laminam subæquante, perigynii foliolis interioribus membranaceis longe unguiculatis cordatis repando-mucronulatis trinervibus margine nudis, culmo oblique tetragono, vaginis superius foliisque hirsutis, foliis intimis brevibus lanceolatis.—*Wight. cat. n.* 1862; *Vahl. en.* 2. *p.* 386; *R. et Sch. s. v.* 2. *p.* 234; *Willd. sp. pl. (ed. Dietr.)* 1. *p.* 257 (excl. syn. *Roxb.* et *Linn.*)—Fuirena, *Wall. cat. n.* 3544. *a.*—Scirpus ciliaris, *Rottb. gr. p.* 55. *t.* 17. *f.* 1.—*Pluk. t.* 417. *f.* 6.——Peninsulæ Indiæ orientalis; *Wight; Rottler; Klein.* China; *Potts.*

Distinctissima species, quantumvis similis *F. pentagonæ.*

TRIB. III. SCIRPEÆ.

XII. DICHOSTYLIS. *Lestib. (ex parte.)*

Spiculæ trifariam imbricata, compressione quandoque ad speciem bifariæ et cyperoideæ. Squamæ omnes fertiles, tenues, membranaceæ et molles, persistentes, carina uninervi in mucronem magis minisve conspicuum prominente. Perigynium nullum. Stylus bifidus, rariusve trifidus, filiformis, ovario continuus diutiusque persistens, demum basi rumpens. Caryopsis oblonga, mucronulata, lævis, distincte marginulata, latere postico plano, antico convexo aut obtuse carinato.—Inflorescentia: spiculæ in capitulum longe involucratum, bracteolis foliaceis aut saltem scariosis et squamiformis interstinctum arctissime confertæ, pallidæ, irregulari ambitu, carina et apicibus virides.

Plantæ humiles *Kyllingiæ* habitu, culmo trigono basi folioso. Folia plana, carinata, longa. Inter *Cyperos* accedit *C. pygmæus;* sed isti spiculæ applanatæ et evidenter distichæ.—Genus *Dichostylis* ad paucas species reduxi, quibus stylus diu in fructu oblongo persistit et quarum inflorescentia capitata, spiculæ parvæ difformes flaccidulæ e squamis plerumque trifariis, habitus omnino Kyllingioideus, differentiam insignem præbent.

1. D. *Micheliana* (N. ab E.:) spiculis multifloris, squamis apice longe subulatis, capitulo bracteis foliaceis interstincto, involucro 4–6-phyllo patentissimo.—Isolepis Micheliana, *R. et Sch. s. v.* 2. *p.* 114; *N. ab E. in Royle. cyp. n.* 92.—Scirpus Michelianus, *Linn. sp. pl. p.* 67; *Willd. sp. pl.* 1. *p.* 309 (excl. syn. *Bauh.*); *Schrad. fl. Germ.* 1. *p.* 147; *Schkuhr in Ust. ann.* 1. *p.* 20. *t.* 2; *Host. gr. austr.* 3. *t.* 69.—*Link. H. Ber.* 1. *p.* 303.——Nepalia; *Royle* (forma pumila). Europa australis et Asia Minor.

Hanc ne cum *Cypero pumilo* confundas, cave. Quod accidisse videtur præcl. Linkio, cum in hort. Ber. 1. p. 303 hæc traderet: *Cyperus,* inquit, *pumilus* ex India orientali multo minor quidem et omnibus partibus rigidior (hæc in verum cadunt *C. pumilum*) at specie differe a *C. Micheliano* non videtur.

XIII. ABILDGAARDIA. *Vahl.* (ex emmend. *R. Br.*)

1. A. *Rottboelliana* (N. ab E. :) diandra, spica solitaria nuda subnutante ovata, squamis duabus inferioribus longius mucronatis reliquis conformibus, caryopsi argute tuberculata, culmo compresso-angulato gracili, foliis culmo brevioribus anguste linearibus obtusiusculis margine scabris.—A. Indica, *N. et E. in cyp. Royl. n.* 10.—α; culmo fere pedali.—*Wight. cat. n.* 1863.—Abildgaardia, *Wall. cat. n.* 3491.—Cyperus monostachyus, *Rottb. gram. p.* 18. *t.* 13. *f.* 3.—Scirpus schœnoides, *Koen.; Roxb. fl. Ind.* 1. *p.* 221; (*ed. Wall.*) 1. *p.* 223 (excl. syn. *Retz.*); *in cœt. merc. Ind. or. mus. tab.* 750 (fide *Arn.*)—β; culmo humiliore, *Wight. cat. n.* 1863. *b.*——In pascuis siccis Peninsulæ Indiæ orientalis, inter frutices; *Roxburgh; Wight.*—β. In Nepalia; *Royle.* Peninsula; *Wight.*

A. compressa, Presl., vic ac ne vix quidem differt tuberculis fructus minoribus acutis nec obtusis. Folia in majoribus angustiora, in minoribus pro longitudine paulo latiora. *A. Indica*, Vahl., quæ *Cyperus Indicus*, Rich., Pers. syn. 1. p. 65, culmo stricto, " monandra," stylo bifido, etc., aliena certo a nostra est species. *A. compressa*, Presl., sic circumscribenda: culmo compresso, foliis subcanaliculatis angustissimis margine scabris culmum subæquantibus, spicula solitaria (rarissima spiculis geminis) ovata, glumis ovato-subrotundis inferioribus mucronatis, caryopsi obovata grosse tuberculata.

2. A. *tristachya* (Vahl. :) spicis triandris subternis, media sessili, lateralibus longe pedunculatis, squamis superioribus spicatis, involucro 1–2-phyllo umbella breviori, caryopsi compresso-trigona angulis lateralibus argutis denticulatis.—*Vahl. en.* 2. *p.* 297 (excl. syn. Schœni cyperoidis, *Retz.* ?); *R. et Sch. s. v.* 2. *p.* 160; *Wight. cat. n.* 1864.—Cyperus triflorus, *Linn. mant. p.* 180; *Willd. sp. p.* 1. 872.——Peninsula Indiæ orientalis; *Rottler; Wight.*

Squammæ revera distichæ, torsione rachillæ quadrantem circuli eminente in duas abeunt spiras. *Schœnus cyperoides*, Retz., cui spiculæ omnes sessiles, distinctissima esse videtur species.

3. A. *Eragrostis* (N. et Meyen :) culmo angulato subcompresso, foliis linearibus obtusis margine serrulato scabris glaucis, umbella decomposita radiolis patentibus, involucro et involucellis brevissimis linearibus, spiculis multifloris pallidis, squamis lævibus acutis, stylo trifido.—*N. et Meyen in act. acad. nat. cur.* 16. *suppl.* 2.——In China; *Potts; Meyen.*

4. *A. fusca* (N. ab E. :) culmo filiformi subtetragono sulcato, foliis linearibus obtusis margine serrulato-scabris viridibus, umbella composita, involucris et involucellis subulatis brevibus, spiculis multifloris ferrugineo-fuscis, squamis acuminatis subtiliter nervoso-striatis scabris, stylo tridentato.—Fimbristylis, *Wall. cat. n.* 3530.——Nepalia; *Wallich.*

Differt ab *A. Eragrosti*: culmi graciliori, foliis culmo multo-brevioribus, spiculis ferrugineo-fuscis, nec pallidis, squamis nervoso-striatis et scabris tum inprimis stylo apice non fisso sed dentibus tribus terminalibus predito.—Binis his speciebus umbellifloris proxima et certo ejusdem generis est: *A. cyperoides*, culmo triquetro hinc sulcato, foliis linearibus obtusis margine scabris, umbellæ decomdositæ radiis pubescentibus, involucro et involucellis umbella umbellulisque brevioribus, spiculis subbifloris.—*N. et Meyen. l. c.*—Gussonea cyperoides, *Presl. in rel. Hænk.* 3. *p.* 183. *t.* 33.——Monte Rey in California; *Hænke.*

XIV. CHÆTOCYPERUS. *N. ab E.*

Spicula solitaria, disticha, pauciflora. Perigynii setæ sex, cartilagineæ, retrorsum hispidæ. Stamina tria. Stylus trifidus, basi bulbosa ovario articulo inserta persistente. Caryopsis trigona, styli bulbo brevi trigono-conico basi trilobo coronata.

A *Limnochari* differt spicula disticha et rostro caryopseos basi trilobo.

1. C. *Limnocharis* (N. ab E.:) spica ovata.—Cyperus setaceus, *Retz. obs.* 5. *p.* 10; *Willd. sp. pl.* 1. *p.* 129; *Roxb. fl. Ind.* 1. *p.* 185; (*ed. Wall.*) 1. *p.* 190.—Eleocharis chaetaria, *R. et Sch. s. v.* 2. *p.* 154.—Scirpus pygmaeus, *Lam. ill. gen.* 1. *p.* 139; *Vahl. en.* 2. *p.* 242; *R. et Sch. s. v.* 2. *p.* 122.—— Coromandel; *Roxburgh.* Ceylona; *Macrae.*

Caryopsis obovata, trigona, fusca, grosse confertimque punctato-striata. Rostrum pyramidale, trigonum, basi dilatata in singulo caryopseos angulo lobo obtuso incumbens. Setæ lutescentes, 4 caryopsin æquantes, duæ paulo breviores. Culmi basi procumbentes, flaccidi, collabentes, prolifero-fasciculati, tenerrimi, capillares, profunde sulcati.

XV. ECHINOLYTRUM. *Desv.*

Est genus plantarum singulare, Fimbristylin omnino quod ad characteres adeo referens, ut diu hæsitarem, monstrosam genuinæ cujusdam Fimbristyleos, maximeque *F. æstivalis* prolem hanc esse censerem. Sed semper et ubique in plurimis spiculis eadem occurrebat ovarii et fructus structura, ut diutius dubitare non liceret. Ovarium e membrana contextum est tenui pellucida, fibris (spiralibus uti videbatur) in areolas transversales elongatas irregulares divisa. Ad latus utrumque, rarius etiam in dorso, series posita est appendicum, 6–8-narum in singula serie, ratione ovarii magnarum, globosarum, brevi crassoque stipiti impositarum. Capitula hæc (uti et stipites) structuræ sunt laxe cellulosæ, cellulis etiam fibrosis. Semen obscurum, basi paulo latius, ubi raphe a stylo decurrens inseritur, ad apicem fere ovarii protensum, denique cavum omne ovarii implens. Pupam dipteri cujusdam, appendicibus sordibusve inquinatam, primum adspicienti ejusmodi caryopsis apprime fingit.

1. E. *dipsaceum* (Desv.)—*Desv. journ. de bot.* 1. *p.* 20. *t.* 1; *Wight. cat. n.* 1865.—Isolepis dipsacea, *Vahl. en.* 2. *p.* 276; *R. et Sch. s. v.* 2. 119.—Scirpus dipsaceus, *Rottb. gram. p.* 56. *t.* 12. *f.* 1; *Willd. sp. pl.* 1. 1. *p.* 309.—— Peninsula Indiæ orientalis.

Stamen unum offendi in nostris, anthera lanceolata lutea.

XVI. FIMBRISTYLIS. *Vahl.*

§ 1. *Spiculis 1–4, grandiusculis non aut breviter involucratis pallidis; caryopsi nodulosa aut striolata, aut cingulata.*—Oligostachyæ.

1. F. *acuminata* (Vahl.:) spicula solitaria ovali-oblonga acuta pallida, squamis ovatis acutiusculis paucinervibus triandris, involucro nullo, caryopsi late obovata biconvexa transversim undulato-cingulata, culmo tetragono-compresso vagina aphylla oblique truncata.—*Vahl.* 2. *p.* 285; *R. et Sch. s. v.* 2. *p.* 91.—Isolepis, *Wall. cat. n.* 3494.——Kamaon; *Wallich.* Ceylon; *Macrae.*

2. F. *nutans* (Vahl.:) spicula solitaria ovali nutante, squamis ovato-subrotundis mucronulatis chartaceis dorso tenuiter nervosis latere fusvescentibus triandris, involucelli loco squama inferiore longius mucronata persistente sterili, caryopsi obovata plano-convexa transversim remote noduloso-rugosa (varicosa), culmo tetragono-compresso ($\frac{1}{2}$–1 ped. alto) aphyllo, vaginis oblique truncatis.—*Vahl. en.* 2. *p.* 285; *R. et Sch. s. v.* 2. *p.* 91.——Ceylona; *Macrae.* Malacca; *Koenig.*

3. F. *polytrichoides* (Vahl.:) spicula solitaria oblonga subnutante, involucro monophyllo e basi membranacea setaceo spicula breviori, squamis membranaceis uninervibus ovato-oblongis mono-diandris, inferioribus subulato-mucronatis superioribus obtusis, stylo inferiorum trifido superiorum bifido scabro bulbo angusto, caryopsi obovata dorso convexa punctulato-scabra, inferiorum costa dorsali media prædita, foliis setaceis compressis complicato-caniculatis lævibus culmo setaceo compressiusculo plerumque brevioribus.—*Vahl. en.* 2.

p. 249; *R. et Sch. s. v.* 2. *p.* 101; *mant.* 2. *p.* 59; *Brown. prod.* 1. *p.* 226?; *Roxb. in cœt. merc. I. o. mus. tab.* 748 (fide *Arn.*); *Wight. cat. n.* 1866.—Scirpus polytrichoides, *Retz. obs.* 4. *p.* 11. *n.* 21; *Willd. sp. pl.* 1. *p.* 295.—S. scaber, *Roxb. fl. Ind.* 1. *p.* 220; (*ed. Wall.*) 1. *p.* 222.—Isolepis, *Wall. cat. n.* 3487, *c.*——Peninsula Indiæ orientalis, ad oras Coromandel in pascuis arenosis siccis; *Roxburgh; Wight.* In Ceylona; *Koenig.*

F. polytrichoidem, R. Br. propter caryopsin "lævem" ab eo descriptam, dubuis citavi. Vis styli bi-vel trifidi in generibus distinguendis hoc exemplo sane infirmatur, nec omnino tamen frangitur, cum restet caryopseos figura plano-convexa, costa illa quidem angulum moliente nec vero perficiente, quo sequitur, Cyperaceas ad fructum omnes tricarpellarem natura sua natas, alias ejusmodi fructum facile perficere posse, alias autem difficile et difficillime. Atqui genus est gradus quidam perfectionis nisus communis familiæ cujusque, plantis quibusdam e ditione ejus collatus, quo adepto nisus ad altiora hac ex parte sedatur, nisi obstet aliquid stimuletve; ergo et confirmatur regula nostra isto exemplo.

4. F. *bispicata* (N. et Mey.:) spiculis 1–3 ovali-oblongis specie lateralibus una sessili reliquis pedicellatis, involucello proprio brevissimo, squamis subrotundis mucronatis, caryopsi pedicellata lenticulari, stylo bifido, culmo filiformi incurvo basi folioso.—α; spica singula vel gemina, involucro majori (ubi altera adest spicula) pedicellum æquante.—Scirpus bispicatus, *Koenig in Roxb. fl. Ind.* 1. *p.* 220; (*ed. Wall.*) 1. *p.* 223; *Spr. s. v.* 1. *p.* 212.—Isolepis bispicata, *Schult. mant.* 2. *p.* 61.—β; spicis geminis ternis quaternisve, involucro majore spiculas superante.—Scirpus tristachyos, *Roxb. fl. Ind.* 1. *p.* 221; (*ed. Wall.*) 1. *p.* 224.—Isolepis tristachya, *Sch. mant.* 2. *p.* 64 (quoad syn. *Roxb.*)——α; In China; *Meyen; Vachell.* In Peninsula Ind. or. et Benghala; *Koenig; Roxburgh.*——β; In Benghalia; *Roxburgh.*

[*Scirpus tristachyus*, Roxb., cui spicæ 3–4 subsessiles aut, ex tab. picta (in cœt. merc. I. o. mus. t. 1931), omnes plane sessiles et aggregatæ, vix hujus loci; (conf. *F. ferugineam*). In *Sc. bispicato*, Roxb. (tab. 749) delineatur inflorescentia trispicata, spica una sessili cæteris pedunculatis, et species omnino eadam ac Wight. cat. n. 1867 (herb. Wight. ad Nees. sub. no. 100), ad *F. schœnoidem* ex Neesio referenda.—ARN.]

5. F. *schœnoides* (Vahl.:) spiculis subgeminis ternis quaternisve ovatis obtusis una sessili, involucro nullo, squamis subrotundis mucronatis, caryopsi pedicellata obovato-lenticulari confertim granulato-striata, culmo filiformi basi folioso, foliis lineari-filiformibus canaliculatis margine scaberrimis.—*Vahl. en* 2. *p.* 286; *R. et Sch. s. v.* 2. *p.* 93; *N. ab E. in herb. Royle. n.* 70; *Wight. cat. n.* 1867.—β, *Monostachya;* spicula solitaria basi bibracteolata, bracteis e latiore basi subulatis, culmo humiliore 2–6 poll. longo, rhizomate repente.—*Wight. cat. n.* 1868.—Scirpus schænoides, *Retz.* ? *obs.* 5. *p.* 14.—S. monostachys, *Koen. in Roxb. fl. Ind.* 1. *p.* 19; (*ed. Wall.*) 1. *p.* 223; *in cœt. merc. Ind. or. mus. tab.* 747 (fide *Arn.*); *Wall. cat n.* 3490.——Peninsula Ind. or.; *Koenig; Roxburgh; Wight.* Nepala; *Wallich; Royle.*

[Exempla Wightiana, No. 1867, ipsissima sunt ac *Scirpus bispicatus*, Roxb., et involucro saltem 1-phyllo (folio pedicellos fere æquante unciam longo) gaudent: quandoquidem autem casu vel ætate tantum involucrum deest.—ARN.]

§ 2. ***Spiculis teretibus monostachyis squamis dorso haud alte carinatis nec acuminatis aut longe mucronatis aristatisve, sæpe obtusis umbellatis, umbella aut simplici, aut composita, et tum vero spiculis non capitatis, sed pedicellatis cum centrali sessili.***—Umbellares.

a. *Umbella simplici.*

6. F. *ferruginea* (Vahl:) umbella brevi simplici (vel rarissime subcomposi-

ta—ARN.) oligostachya, spiculis ovali-oblongis acutis, squamis ovatis acutiusculis dorso supero sericeo-tomentosis ferrugineis triandris, caryopsi obovata marginata subtiliterque punctato-striata matura fusca, involucro diphyllo stricto umbella breviore paulove longiove, culmo compresso subtrigonove lævi glauco, foliis convoluto-linearibus margine spinuloso-ciliatis culmo multo brevioribus [quandoque subnullis—ARN.], vaginis truncatis ferrugineis ore ciliatis.—α; umbella tri-hexastachya involucro breviori, culmo obtuse trigono. —F. ferruginea, *Vahl. en.* 2. *p.* 291.—Schœnus polymorphus, *Rottb. gram. p.* 97; *R. et Sch. s. v.* 2. *p.* 132.—β; umbella mono-hexastachya involucro longiore.—*Wight. cat. n.* 1869.—F. ferruginea, *Sieb. fl. Maur.* ii. *n.* 24.—F. arvensis, *Vahl. en.* 2. *p.* 291; *R. et Sch. s. v.* 2. *p.* 99.—Scirpus arvensis, *Retz. observ.* 4. *p.* 11 (non *Roxb.*).—S. Tranquebariensis, *Roth. cat.* 3. *p.* 6; *nov. pl. sp. p.* 29; *R. et Sch. s. v.* 2. *p.* 137.—γ, *Glomerata;* spiculis paucis agglomeratis involucro diphyllo multo brevioribus, culmo compresso.—*F.* compressa, *R. et Sch. s. v.* 2. *p.* 100.—Scirpus Bonariensis, *Poir. enc. meth.* 6. *p.* 763.—[S. tristachyus, *Roxb. fl. Ind.* 1. *p.* 221; (*ed Wall.*) 1. *p.* 224; *in cat. merc. I. or. mus. tab.* 1931.—ARN.]——α. Insula St Vincentii; *Guilding.*—β. In Peninsula Ind. or.; *Wight; Rottler; Klein.* In Ins. St Maurit.; *Sieber.*—γ. Bahiæ in Maritimis. Bonariæ. Beughalia; *Roxburgh.*

Species tomentulo squamarum et caryopsi confertim striolata, primum alba dein fusca, bene distincta. Var. γ differt a communi forma culmo humiliori, irregulariter subangulato, glaucedine, tum spiculis non umbellatis, sed 4–7 aliis sessilibus, aliis brevissime pedicellatis veluti in capitulum collectis, quæ causa videtur, ut involucrum (alterum ½, alterum 1 fere pollicem longum) altitudinem florum magis minusve superet. Spicularum autem structura color et leve tomentum omnino quale in reliquis.

b. *Umbella composita.*

7. F. *podocarpa* (N. et Mey.:) umbella composita pauciradiata contracta, radiis mono-pentastachyis, spiculis ovato-conicis, squamis mono-di-tri-andris subrotundis obtusis brevissime mucronatis, caryopsi distincte stipitata obovata compressa obtuse marginata crenato-costulata serie punctorum interjecta, culmis strictis compresso-angulatis striatis lævibus, foliis linearibus obtusis culmo brevioribus glabris ciliatisve, involucro di-tetraphyllo, altero umbellam superante.—α; glauca, foliis subtus margineque puberulis, radiis 2–3-stachyis.—β; glabra, foliis minus glaucescentibus parum ciliatis involucris hirsutis, radiis 1–2-stachyis.—*Wight. cat. n.* 1870; *Royle. cyp. n.* 72. *et* 77.—F. pilosa β. *Presl. rel. Haenk.* 3. *n.* 191.—γ; major, foliis brevioribus cum vaginis pubescentibus, radiis 4–5-stachyis, involucro tetraphyllo,—*Royle cyp. n.* 68.——α. In China; *Meyen.*—β. In Manilla; *Meyen.* Nepalia; *Royle.* Peninsula Indiæ orient.; *Wight.*—γ. In Nepalia; *Royle.*

Structuram fructus quod attinet, similis isti sola inter cognitas est *F. cincta* (N. ab E.:) umbella subcomposita pauciradiata contracta, radiis monotristachyis, spiculis ovatis, squamis monandris subrotundis, infimis brevissime mucronulatis glabris, caryopsi obovata obtuse marginata punctato-striata, culmis filiformibus compresso-angulatis striatis superne scaberrimis, foliis angustissime linearibus culmum subæquantibus, vaginis involucrisque diphyllis umbella brevioribus glabris margine scabris.—F. marginata, *N. ab E. in Sieb. herb. Maur.* ii. *n.* 22 (non *La Bill.*).

8. F. *ovalis* (N. ab E.:) umbella composita vel subcomposita patula brevi, radiis 1–5-stachyis, spiculis ovalibus glabris, squamis triandris ovatis obtusis rufo-brunneis uninervibus margine albis, caryopsi obovata subsessili alba convexo-plana costata, costis approximatis crenulatis interstitiis subtilissime transversim striolatis, marginibus obtuse denticulatis, involucro diphyllo umbella multo breviore, foliis subradicalibus linearibus oblique acutatis marginatis supra pilosulis, vaginis villoso-sericeis, culmo gracili compresso striato-angulato.—α; minor, umbellæ radiis 1–2-stachyis.—*Wight. cat. n.* 1871.—β;

major, umbellæ radiis 2–5-stachyis.—Fimbristylis, *Wall. cat. n.* 3503 *et* 3509. —Scirpus dichotomus, *herb. Roxb.* (fide *Wall.*)——Peninsula Ind. or.; *Wight.* —β. Nepal, Silhet, Benghalia.

Differt a *F. podocarpa* et *rigidula* fructu minore, nec subpedicellato, nec leniter striato sed costis novem prædito.

9. F. *rigidula* (N. ab E.:) umbella composita patula, radiis 1–3-floris, spiculis ovatis glabris nitidis castaneis, squamis triandris ovatis acutiusculis subtilissime nervoso-lineatis, caryopsi obovata plano-convexa pallida leviter striata tuberculisque sparsis exasperata, involucro diphyllo altero umbellam subæquante, culmo compressiusculo sulcato, foliis subradicalibus linearibus obtusis.—*N. ab E. in Royle. herb. n.* 53, 60, 74.—Fimbristylis, *Wall. cat. n.* 3519.——Nepalia; *Wallich, Royle.*

Similis *F. arvensi,* a qua differt staminibus ternis, nec binis, et caryopsi pallida non sulcata sed leviter striata punctulisque elevatis exasperata. Radiorum unus aut alter distachyus, aut tristachyus. Variat culmo aliquot pollicum longitudine, basi bulboso. An var. *F. arvensis?*

[In exemplis meis caryopsis albida sulcata haud tuberculata, sulcis subcancellatis, omnino ut in Wight. cat. n. 1873.—ARN.]

10. F. *Wightiana* (N. ab E.:) umbella decomposita rotundata densa, umbellulis plurifloris compactis, involucro diphyllo umbella multo breviore, spiculis ellipticis obtusis glabris (parvis), squamis arcte imbricatis late ovatis obtusissimis testaceis apice membranaceo-candidis dorso subtilissime trinerviis diandris, stylo imberbi, caryopsi obovata plano-convexa fusca tuberculis parvis aspera, foliis linearibus obtusiusculis rigidis canaliculatis glabris margine scabris culmo 1–1¼ pedali basi sulcato-angulato apice compresso brevioribus.—*Wight. cat. n.* 1872.—β; umbella fere capitata.—*Wight. cat. n.* 1872. *b.* ——Peninsula Indiæ orientalis; *Klein; Wight.*

Pulchra species, e notis adlatis facile distinguenda, nisi cum *Trichelostyli obtusifolia,* fugaci examine instituto, composueris. A *F. Torresiana,* Gaudich., jam eo differre videtur quod sit *Fimbristylis,* neque *Dichostylis;* taceatur etiam l. c. folia et fructus; nec culmus teres est, sed sulcatus.

[Ad α verosimiliter pertinet *S. strictus,* Roxb. (in cœt. merc. I. o. mus. tab. 756), habitu characteribusque satis congruens.—ARN.]

11. F. *Royeniana* (N. ab E.:) umbella composita decompositaque brevi densa, umbellulis plurifloris radiolis patentibus, involucri diphylli foliolo altero umbella breviore paullove longiori, spiculis ovatis obtusis cum mucronulo membranaceis pallide testaceis dorso tenuissime trinervibus di-triandris, caryopsi obovata basi acutata obtuse marginata utrinque leviter quinquecostata sulcis planis transversim striatis, foliis anguste linearibus obtusis culmo compresso-trigono brevioribus margine scabriusculis, vaginis inferioribus lanuginosis.—α; umbella pauciradiata contracta, culmo gracili, vaginis glabriusculis ore tantum pubescentibus.—Scirpus dichotomus, *herb. Royeni.*—β; umbella magis patente atque composita triradiata, vaginis densius lanuginosis. —*Wight. cat. n.* 1873? *et* 1874.—Fimbristylis, *Wall. cat. n.* 3505. *c. f.*—γ, *Microstachya;* culmo radiisque umbellæ elongatis, spiculis minoribus, demum ex oblongo-lanceolatis semper triandris, caryopsi altius costata.—*Wight. cat. n.* 1875.—F. glauca, *Vahl. en.* 2. *p.* 218; *R. et Sch. s. v.* 2. *p.* 96.—γ, *Procumbens;* culmis (2–4 poll. longis) procumbentibus.——Peninsula Indiæ or.; *Wight.* Benghalia inferiore; *Roxburgh.* Ceylona.

Var. γ figura spicularum accedit ad *F. diphyllam,* sed differt spiculis et præsertim caryopsi minoribus. Involucrum in hac quandoque umbellam superat. Umbella sæpe decomposita, radiis pluribus di-tristachis. δ (in herb. Wightiano) non differt, nisi culmis ad horizontem [exsiccatione et arte tantum—ARN.] expansis: radix in hoc nostro exemplo constat e barba fibrarum longissimarum spongioso-mollium patenti-hirsutarum.

[Wall. cat n. 3509. *c,* vel Wight. cat. n. 1873, huc a Neesio ad var. β trahitur, quanquam achenium multo majus et utrinque 9-costatum sulcis specie cancellatis ut in *F. rigidula* et *F. diphylla*, a qua differe videtur staminibus 2–3.—Var. δ solummodo dignoscitur culmo humiliore; itaque potius ad β repellenda.—*S. diphyllus*, Roxb. (*S. fuscus*, R. in cœt. merc. I. o. mus. tab. 758) ad β accedit habitu omnibusque notis una tantum excepta, caryopsi "smooth, brown," et in tab. picta lævi delineata, forsan per incuriam.—ARN.]

12. **F. *tomentosa*** (Vahl:) umbella subdecomposita contracta, spiculis in radiolis solitariis ovatis, squamis acutis monandris, caryopsi cancellata, culmis angulatis sulcatis glabris, foliis linearibus culmum subæquantibus pilosis, involucro tri-pentaphyllo, foliolo uno alterove umbellam superante æquanteve, invollucellis brevibus setaceis.—*Vahl. en.* 2. *p.* 290; *R. et Sch. s. v.* 2. *p.* 97.——In India orientali. In China, et in Manila; *Meyen*. In Java insula; *Blume*.

Apud Vahlium, l. c. pro "apice" certo legendum "basi" (scil culmi) 2–3-phylli. A *F. elongata*, R. Br., imprimis differt involucellis brevibus et stylo apice nec basi ciliato.

13. **F. *diphylla*** (Vahl:) umbella decomposita laxa divergente, spiculis in radiolis solitariis ovato-lanceolatis acutis, squamis acutis monandris, caryopsi obovata brevi cancellata, culmis angulatis sulcatis foliisque linearibus margine serrulato-scabris glaucis curvisque glabris, superioribus culmum æquantibus, involucro di-tri-phyllo, foliolo uno umbellam subæquante vel nonnihil superante, involucellis brevibus setaceis.—*Vahl. en.* 2. *p.* 289; *R. et Sch. s. v.* 2. *p.* 97; *Spr. s. v.* 1. *p.* 201.—Scirpus diphyllus, *Retz. obs.* 5. *p.* 15; *Willd. sp. pl.* 1. *p.* 301.—S. miliaceus, *Roxb. fl. Ind.* 1. *p.* 227; (*ed. Wall.*) 1. *p.* 231. —*Pluk. t.* 199. *f.* 3.——In India orientali; *Koenig*. In China; *Meyen*.

Simillima *F. tomentosæ* (cujus forsan varietas) differt: foliis glabris, angustioribus, juxta basin culmi uno duobusve, margine præsertim circa apicem obtusum serrulatis infimis sulcatis; vaginis glabris, membrana fusca hirtula; umbellæ radiis plerisque magis elongatis, longiori ultra-tri-pollicari, latioribus; spiculis longioribus, 3–4 linearum; squamis palide nec saturate fuscis, carina viridi; caryopsi obovata, magis compressa, matura lutescente et denique fusca, in latere exteriori magisque convexulo costis septem, in interiori novem prædita, sulcis interjectis simplici serie grosse punctatis porcatis termino entomologico. Dum caryopsis statu immaturo siccatur, pellicula alba Margaritarum ad instar nitente vestita apparet et tum vero in longitudinem et transversim undatim sulcata haud male eam cum Vahlio describes. *Scirpus diphyllus*, Roxb. fl. Ind. longe diversa est species.

[*S. miliaceus*, Roxb. (tab. pict. n. 759, sub *S. glomerato*), certe ad *F. Royenianam* γ repellendus.—ARN.]

§ 3. *Spiculis teretibus capitatis.*—Capitatæ.

14. **F. *albicans*** (N. ab E.:) spiculis capitatis oblongis, squamis membranaceis oblongis acutis monandris, caryopsi obovata transversim tuberculato-undata, involucro di-triphyllo capitulo subduplo longiore, foliis radicalibus linearibus obtusis planis recurvis culmo compresso brevioribus, culmeo nullo. —*Wight. cat. a.* 1877.——Peninsula Indiæ orientalis.

Distincta a *Fimbristyli argentea*: culmo ½–¾ ped. alto, sat lato, compresso, foliis lineam fere latis obtusis nec subulatis.

[Quanquam non vidi Wall. L. n. 3482, satis certus sum hanc esse speciem. —ARN.]

15. **F. *argentea*** (Vahl:) spiculis capitatis subcylindricis, squamis monandris-diandris ovatis submucronatis dorso viridi-albicantibus trinervibus lateribus e ferrugineo pallescentibus, rhachi alveolata, caryopsi obovata subtilissime transversim rimulosa luteola, involucro tri-tetraphyllo capitulo triplo saltem longiore, foliis linearibus complicato-setaceis obtusiusculis margine scabris, culmo trigono-compresso basi tantum folioso.—*Vahl. en.* 2. *p.* 294; *R. et Sch.*

s. v. 2. *p.* 101; *Wight. cat. n.* 1878.—Scirpus monander, *Rottb. gram. p.* 50. *t. f.* 3; *Willd.? sp. pl.* 1. *p.* 511.—S. argenteus, *Roxb. fl. Ind.* 1. *p.* 223; (*ed. Wall.*) 1. *p.* 226; *in cæt. merc. Ind. or. mus. tab.* 754 (ex *Arn.*); *Willd. sp. pl.* 1. *p.* 311.—Isolepis, *Wall. cat. n.* 3183.——In Peninsulæ Ind. or. pascuis arenosis, denso cæspite crescens, culmis periphericis sursum incurvis.

Est species distinctissima, sed synonymis dubiis vexata. Differt a *F. albicante* culmo humili, 1–6 poll. longo, foliis angustioribus, spiculis longioribus, squamis firmioribus dorso viridibus trinervibus ad latus nervorum fusco-rufis, circa margines pallidis, caryopsi exigua, fere lævi. Stamen unum, nec plura inveni, ideoque *S. argenteum*, Retzii, cum dubitatione adduxi. Singularis est rachillæ structura, quæ alveolis 5-angularibus profundis membranaceis prædita est, ad fructus ample fovenda. Styli bifidi subciliati bulbus crassus, ovatus.

§ 4. *Spiculis propter squamas altius carinatas, sæpe etiam acutas, angulatis.*

a. *Spiculis capitatis, capitulis umbellatis.*

16. F. *decora* (N. et Meyen:) umbella composita, spiculis capitatis ovato-oblongis, squamis acutis triandris, stylis bifidis, culmis striatis strigiloso-puberulis, foliis culmo brevioribus acutis supra argenteo-sericeis, involucro brevissimo.——In China; *Meyen, Potts.*

Similis *F. sericeæ*, R. Br., a qua differt stylo bifido et foliis supra nec subtus argenteis.

b. *Spiculis umbellatim dispositis.*

17. F. *pallescens* (N. ab E.:) umbella decomposita et supradecomposita involucro diphyllo breviori, spiculis (parvis) ovato-oblongis acutis, squamis ovatis obtusis cum mucronulo pallide testaceis albo-marginatis monandris, caryopsi obovata brevi utrinque quinque-costata sulcis transversim striatis, foliis anguste linearibus acuminatis involucrisque scabris aut pubescenti-tomentosis incanis.—α; foliis involucrisque villosis, culmo hirto.—*Wight. cat. n.* 1879.—Fimbristylis, *Wall. cat. n.* 3515. *b*, 3516. *b.*—Scirpus dichotomus, *Rottb. gr. p.* 57. *t.* 13. *f.* 1.—S. dichotomus β. villosa, *R. et Sch. s. v.* 2. *p.* 95.—S. bisumbellatus, *Forsk. descr. p.* 15.—S. pallescens, *Roxb. fl. Ind.* 1. *p.* 229; (*ed. Wall.*) 1. *p.* 233.—β; foliis involucrisque scabris aut omnino glabris.—F. dichotoma, *Vahl. en.* 2. *p.* 287; *R. et Sch. s. v.* 2. *p.* 95; *N. ab E. in herb. Royle. n.* 45–47.—Scirpus dichotomous, *Linn.? fl. Zeyl. p.* 40.—γ; culmo altiore, foliis sparsim villosulis, umbella magis composita et divaricata.—*Wight. cat. n.* 1879. *b.*—F. dichotoma, *N. ab. E. in herb. Royle.*——Peninsula Ind. or.; *Roxburgh; Rottler; Wight.* Benghalia inf. et Nepala; *Wallich; Royle.*

Synonyma, quæ supra attuli certiora existimo, Linnæana omnia autem maxime incerta. E flora Ceylanica vidi exemplum herb. Royeni, *S. dichotomo* nomine inscripto, et probabiliter a Linnæo olim comprobatum, quod diversæ est speciei, ejusque ad *F. diphyllam*, Vahl., propius accedentis sed, quod ad fructum attinet, *Fimbristyli tomentosæ* propinquioris. Icon Pluken. t. 119. f. 3, quam in sp. pl. 1. p. 74. solam hujus speciei testem adhibet Linnæus, in formas quasdam *F. diphyllæ* congruit, minime autem in nostram. Haud improbabile igitur videtur, Linnæum *F. diphyllam* potius aut formam magis explicatam *F. Royenianæ* nostræ, quam ejus, quæ hodie *dichotoma* appellari solet, hoc nomine subintellexisse.

[*S. brevifolius*, Roxb. (tab. pict. 756 sub *S. autumnali*,) huc referri debet: at *S. pallescens*, Roxb. (tab. pict. 763), multo magis dubius ob squamas obtusas et, ex tabula, minime mucronatas, affinior videtur *F. diphyllæ*.—ARN.]

18. F. *dichotoma* (Vahl:) umbella subcomposita radiis tortuoso-divergentibus, involucro di-triphyllo angusto umbella breviori, spiculis ovato-oblongis (mediocribus), squamis ovatis patulo-acutis carinatis dorso trinervibus latere luteis monandris, caryopsi obovata lævissima, foliis lineari-filiformibus culmo

brevioribus involucrisque pubescentibus.—*Vahl. en.* 2. *p.* 287 (var. *a.*); *R. et Sch. s. v.* 2. *p.* 95; *Link. h. Ber.* 1. *p.* 290.—Fimbristylis, *Wall. cat. n.* 3516. *a* (ex parte).——In Nepalia; *Wallich.*

Similis est *Fimbristyli pubescenti*, sed abunde diversa caryopsi longiori (ratione latitudinis habita) brevissimaque, nec costata, umbella minus composita, et singulari modo quasi torsione quodam radiorum patens atque depressa. A nostrate *F. dichotoma*, hæcce differt caryopsi vel sub fortissima lente lævi, nec omnino striata. An species distincta? An *F. æstivalis* var. spiculis majoribus squamisque minus cuspidatis?

19. F. *æstivalis* (Vahl:) umbella composita subdecompositaque radiis strictis plurifloris, spiculis ovato-oblongis acutis, squamis ovatis mucronato-acuminatis arista recurva uninervibus monandris, caryopsi obovata biconvexa lævi pallida, involucri 2–6-phylli foliolis tribus umbella longioribus, involucellis foliisque lineari-setaceis culmum subæquantibus spiculisque villosis, culmo humili compresso-angulato.—*Vahl. en.* 2. *p.* 288; *R. et Sch. s. v.* 2. *p.* 596; *Wight. cat. n.* 1880.—Fimbristylis, *Wall. cat. n.* 3516. *a* (ex parte), 3517. *c.*—Scirpus æstivalis, *Retz. obs.* 4. *p.* 12; *Willd. sp. pl.* 1. *p.* 398; *Roxb. fl. Ind.* 1. *p.* 227; (*ed. Wall.*) 1. *p.* 230 (excl. observ. *Wall.*).——Peninsula Ind. or. Ceylona. Nepalia; *Roxburgh.* Ripa Irawaddi, *Wallich.*

20. F. *comata* (N. ab E.:) umbella simplici oligostachya involucro diphyllo basi scarioso hirtoque longiori, spiculis oblongis glabris, squamis ovatis aristato-acuminatis acumine recurvo dorso trinervibus monandris, stylo adhærente, caryopsi obovata lævi a pericarpio a basi sursum in fibras soluto calyptrata, foliis filiformibus culmo brevioribus, vaginis hirtis, culmo humili compresso-augulato.—Isolepis, *Wall. cat. n.* 3479, *c.*—Scirpus æstivalis, *Wallich in observ. ad speciem hujus nominis in Roxb. fl. Ind. or.* (*ed. Wall.*) 1. *p.* 230. ——In Benghalia inferiore; Nepalia; Sillet; *Wallich.* Patna et Salpighurri; *Hamilton.*

Distinctissima est species, facilique negotio ab antecedente distinguenda. Culmi humiliores, folia et involucra sunt breviora, umbella 3–4-stachya, spiculis 2–3 pedunculatis mediaque sessili. Stylus basi incrassatus, cum pericarpio arctius conjunctus ejusdemque fibris quasi retentus, superne pilosulus, breviori spatio bifidus.

21. F. *scaberrima* (N. ab E.:) umbella decomposita et supradecomposita involucro diphyllo breviori, spiculis (parvis) ovalibus paucifloris, squamis triandris ovatis obtusis submucronatis alte carinatis carina subtiliter nervoso-striata lateribus membranaceis luteo-nebulosis, foliis longis linearibus planis obtusis glaucis margine facieque apicem versus scaberrimis, culmo compresso ancipiti latere altero plano uninervi margine scaberrimo.—Fimbristylis, *Wall. cat. n.* 3507.——In Silhet; *Wallich.* Peninsula Ind. or. *Heyne.* Mauritii ins.; *Telfair.*

Species hæc primo intuitu persimilis esse videtur *Trichelostyli complanatæ* sed differt non modo stylo (ultra medium) bifido, sed etiam forma spicularum atque squamarum florantium, foliis involucrisque longioribus, illis culmum (pedalem) æquantibus, facie apicem versus scaberrimis magisque glaucis, tum culmi marginibus scindentibus scaberrimis nec læviusculis.

VI. TRICHELOSTYLIS. *Lestiboudois; N. et Mey.*

§ 1. *Spicula solitaria, terminali.*

1. T. *filiformis* (N. ab E.:) culmo filiformi striato, vagina baseos aphylla oblique truncata, spiculæ lanceolatæ squamis ovalibus obtusis dorso striatis duabus intimis majoribus, caryopsi obovata tuberculata.—Isolepis, *Wall. cat. n.* 3488. *b.*——Singapur; *Wallich.*

Fimbristylin acuminatam inter *Trichelostyles* refert, sed habitu omnino recedit ab hac et a reliquis, presertim quod ad squamas spiculæ attinet.

§ 2. *Spiculis umbellatis, radiolis extremis 2–3-stachyis, spiculis lateralibus distincte pedicellatis.*

a. *Culmo compresso, aut ancipiti, aut trigono.*

2. T. *tenella* (N. ab E.:) culmo filiformi angulato compressiusculo, umbella subcomposita radiis tri-di-monostachyis, spiculis ovato-oblongis media radiorum sessili lateralibus longe pedunculatis patentibus, squamis ovatis acutis cum mucronulo carinatis glanduloso-punctatis brunneis carina viridi trinervi, caryopsi obovata obtuse trigona grosse tuberculata alba, involucro ditriphyllo umbella multo breviore, foliis lineari-filiformibus culmoque apice scabris.—*Wight. cat. n.* 1881.—Scirpus tenellus, *Roxb. fl. Ind.* 1. *p.* 225; (*ed. Wall.*) 1. *p.* 227.——Peninsnla Indiæ orientalis; *Roxburgh; Klein; Wight.*
Proxima *T. scabræ*, sed differt gracilitate, spiculis brevioribus, aliisque notis.

3. T. *scabra* (N. ab E.:) culmo trigono striato subcompresso, umbella subcomposita, spiculis lanceolato-oblongis geminis ternisve longe pedicellatis patentissimis cum intermedia sessili, squamis ovatis obtusis cum mucrone glabris castaneis, caryopsi obovata obtuse trigona obsolete striata punctulataque, involucello diphyllo umbella breviore foliisque lineari-falcatis obtusis scaberrimis.—*N. ab E. in herb. Cyp. Royl. n.* 44.——Nepalia.
Similis *Tr. autumnali*, a qua differt: culmo altiori pedali circiter, trigono nec multum compresso, spiculis majoribus obscure castaneis magis patentibus, squamis latioribus obtusis, foliis valde scabris submarginatis dimidii fere culmi longitudine. *Fimbristylin dichotomam* refert; hæc autem differe stylo bifido, nec trifido.

4. T. *complanata* (N. ab E.:) involucro diphyllo umbella supradecomposita divergente breviore eamve subæquante foliisque planis obtusissimis, caryopsi obiter punctata alba, stipite cyathiformi exiguo luteo.—α; involucri folio altero umbellam æquante.—Scirpus complanatus, *Vahl. en.* 2. *p.* 279; *R. et Sch. s. v.* 2. *p.* 119; *mant.* 2. *p.* 56 (*sub* Fimbristyli autumnali).—β; involucri uno alterove folio umbellam superante.—Fimbristylis, *Wall. cat. n.* 3525. *f.*—F. complanata, *Link. hort. Ber.* 1. *p.* 292 (excl. plerisque synon.)—Isolepis Willdenovii, *R. et Sch. s. v.* 2. *p.* 120.—Scirpus anceps, *Willd. Berl. mag.* 2. *p.* 288. *t.* 8. *f.* 2.—Cyperus complanatus, *Willd. sp. pl.* 1. *p.* 270 (excl. syn. *Retz.*)—γ; involucro umbella breviore-(1); spiculis paucifloris.—*Wight. cat. n.* 1882.—Fimbristylis, *Wall. cat. n.* 3522 *et* 3526.—F. autumnalis *var* Indica, *Roth. nov. sp. pl.* 1. *p.* 26.—Scirpus complanatus *Retz. obs.* 5. *p.* 14; *Roxb. fl. Ind.* 1. *p.* 228; (*ed Wall.*) 1. *p.* 232 (excl. syn. *Vahl.*); *in cæt. merc. Ind. or. mus. tab.* 760 (fide *Arn.*)—γ 2; spiculis multifloris.—*Wight. cat. n.* 1882. *b.*——Crescunt var. α et γ in Peninsula Ind. or.; *Koenig; Roxburgh; Heyne; Klein; Wight.* Sunderbun; Nepal; Ava; Silhet; *Wallich.*—β in Java insula.—γ in Cap. b. sp.; *Ecklon.* China; *Meyen; Vachell.* Acræ; *Don.* Bahiæ; *Salzmann.*
An species plures hic latent, denique distinguendæ? Nostris omnibus involucrum est umbella duplo brevius, obtusissimum. Stylus trifidus, altero ramulo breviori. Descriptio Rothiana, ad stigma usque, quod bifidum descripsit, forte neglecto ramulo minore, acuratissima est. Caryopsis non flavida est, sed alba, neque transversim rugosa, sed punctulata, et circa verticem obtusissimum tuberculata, basi contracta et disco brevissime stipitato, margine acuto, a postico latere magis prominente luteo imposita. *T. autumnalis* describenda: involucro diphyllo umbella composita pauciflora breviore foliisque acutis, caryopsi lævi lutescente angulis obtusis discretis.

5. T. *miliacea* (N. ab E.:) involucro involucellisque diphyllis setaceis brevibus basi vaginantibus, umbella supradecomposita subverticillata, spiculis subglobosis, culmo basi quadrangulari apice compresso ancipiti foliisque li-

neari-ensiformibus acuminatis glabris, caryopsi obovata subtrigona muricata transversim striolata, urceolo hypogyno discoideo.—*Wight. cat. n.* 1883.—Fimbristylis, *Wall. cat. n.* 3500 (excl. *c.*) *et* 3524.—F. miliacea, *Vahl. en.* 2. *p.* 287; *R. et Sch. s. v.* 2. *p.* 93; *mant.* 2. *p.* 63 (excl. syn. *Roxb.*), *et p.* 475; *mant.* 3. *p.* 530; *Spr. syst.* 1. *p.* 202; *Link. h. Ber.* 1. *p.* 292.—F. Benghalensis, *R. et Sch. s. v.* 2. *p.* 94; *mant.* 3. *p.* 530.——Scirpus miliaceus, *Rottb. gram. p.* 57 *n.* 77. *t.* 5. *f.* 2; *Linn. sp. pl. p.* 73; *Willd. sp. pl.* 1. *p.* 305; *Burm. fl. Ind.* 22. *p.* 22. *f.* 2.—S. Benghalensis, *Pers. syn.* 1. *p.* 68.—S. tetragonus, *Poir. enc. meth.* 5. *p.* 567; *suppl.* 5. *p.* 91; *R. et Sch. s. v.* 2. *p.* 138; *Sieb. fl. Maur. n.* 26; *Roxb. fl. Ind.* 1. *p.* 228; (*ed. Wall.*) 1. *p.* 232.—β; minor, foliis culmum 4–6-pollicarem superantibus, umbella minore minusque composita.—γ; depauperata, umbella oligostachya, involucro sæpe monophyllo umbella longiore, culmo humili.——In Ind. orient. pascuis humidis. In Peninsula, Gongachora, Silhet, Ava, Nepal. In Ceylona; *Burmann; Macrae.* In Manila in sylvis humidis; *Meyen.* In Java insula; *Blume.* China; *Meyen.* In Madagascar circa Monplaisir; *Aub. de Petit Thouars.*

Descriptiones Poiretiana et Roxburghiana elegantissimæ. Rottbœllius, culmi non nisi summitatem pingens, prorsusque parte inferiore plantæ carens carinam mediam, in altera culmi parte evidentiorem, angulum esse tertium pressione complanatum existimavit et culmum triquetrum descripsit, qui quidem fons omnium errorum, qui seriori tempore hanc speciem vexavere. Icon Rottbœlliana nodum solvit: "Stamina tria, et stigmata duo" verbis, nec figura expressa sunt. Monandrum nos semper, Roxburghius etiam diandrum reperit. Stigmata adsunt tria.

[*Scirpum tetragonum*, Roxb., ex parte ad *T. miliaceam*, ex parte ad *T. tetragonam* pertinere facile patet tam ex descriptione sua quam ex tabula picta (in cœt. merc. Ind. or. mus. n. 761); nec fere ullum descrimen inter has duas species indicare possum: in una et altera folia disticha et ensiformia, culmus aphyllus vaginatus et acute tetragonus; in hac tamen culmi apex transversim sectus figuram apud mathematicos quadram dictam, in illa rhombum exhibet; rhombus autem in nonnullis exemplis ad quadram ita valde accedit, ut hoc vix ex jussu naturæ descrimen est.—ARN.]

b. *Culmo acute tri-aut quadrangulari, aut tereti, sæpe aphyllo aut foliis nullis nisi radicalibus donato.*

6. T. *tetragona* (N. ab E.:) culmo basi aphyllo profunde quadrangulari angulis acutissimis, umbella composita decompositaque patula, spiculis subglobosis, squamis ovatis obtusis membranaceis glabris pallide testaceis, caryopsi subglobosa obtuse trigona scabra alba, involucro involucellisque diphyllis a basi ovata membranacea setaceis umbella umbellulisque brevioribus, foliis radicalibus anguste ensiformibus distichis culmum æquantibus superantibusve lævibus, vagina ad basin culmi aphylla.—*Wight. cat. n.* 1833, *b.*—Scirpus tetragonus, *Roxb. fl. Ind.* 1. *p.* 228; (*ed Wall.*) 1. *p.* 232; *Wall. cat. n.* 3500, *c.*—β; culmo foliis duplo breviore.—*N. ab E. in herb. Cyp. Royle. n.* 69.—γ; exigua, culmo vix semipollicari foliis multo breviori, spiculis paucis in culmi apice subcapitatis involucro brevioribus.——Peniusula Ind. or.; *Wight.* Silhet; *Wallich.* Nepal; *Royle.* Gongachora; *Hamilton.*

Culmus per se aphyllus, inferne vaginis longis membranaceis vestitus. Folia equitantia, disticha, ensiformia, lineari-attenuata, lævia. Vaginæ margine membranaceæ valde compressæ. Persimilis est *T. quinquangulari*; an var.?

7. T. *quinquangularis* (N. ab E.:) culmo profunde quinquangulari, umbella composita decompositaque patente, radiis longioribus demum mutantibus, spiculis ovalibus, squamis monandris ovatis obtusis mucronatis glabris rufo-ferrugineis, caryopsi parva obovata subtrigona lutescente transversim parallele striata tuberculisque compressis subechinata, involucro diphyllo anguste lineari umbella breviori, involucellis setaceis, foliis anguste linearibus glabris margine scabris culmum subæquantibus.—Scirpus quinquangularis.—*Vahl.*

en. 2. *p.* 279; *R. et Sch. s. v.* 2. *p.* 146; *Roxb.? fl. Ind.* 1. *p.* 229; (*ed. Wall.*) 1. *p.* 233.—S. pentagonus, *Roxb. in cat. merc. Ind. or. mus. tab.* 762 (fide *Arn.*)—S. plantagineus, *Roxb. ex Wall. cat. n.* 3512, *a.*—α; umbella minus composita.—*N. ab E. in herb. Royle. n.* 43.—β; umbella decomposita patente.—*N. ab E. in herb. Royle. n.* 67; *Wight. cat. n.* 1884.—Fimbristylis, *Wall. cat. n.* 3512, *c, d.*—α et β in Nepalia; *Royle.*—β Peninsula Indiæ orient.; *Roxburgh; Wight.* Benghala inferiori; *Wallich.*

8. T. *pentaptera* (N. ab E.:) culmo profunde quinquangulari hirsuto, involucris involucellisque brevibus scariosis aristatis hirtis, umbella subdecomposita, spiculis squamisque fuscis rigidulis ovatis acutis, staminibus ternis, caryopsi obovato-trigona tuberculato-aspera; foliis ....——Ceylona; *Macrae.*

Differt a *Tr. angulari* et *curvifolia*, Schrad., tum a *Sc. quinquangulari*, Vahl., et *pentagona*, Roxb., hirsutie et involucrorum.

9. T. *salbundia* (N. ab E.:) culmo profunde quinquangulari glabro, involucro umbella multo breviori subulato foliolis basi involucellisque scarioso-marginatis, umbella decomposita ovata, spiculis ellipticis, squamis ovatis obtusis dorso trinervibus margine membranaceis coloratis, staminibus ternis, caryopsi truncato-obovata obtuse carinata subtiliter punctato-striata sursumque in ambitu tuberculata, foliis nullis.—Fimbristylis, *Wall. cat. n.* 3499. *et* 3526.—Scirpus salbundius, *Hamilt.*——Burrisoil et Munghir; *Hamilton.* Taong Dong, Nepalia, et Silhet; *Wallich.*

Differt a *Tr. pentagona*, culmo profunde quinquangulari nec obiter pentagono, et floribus triandris, a *Tr* (*Scirpo*) *quinquangulari*, Vahl., magnitudine vix dimidia, vaginis aphyllis, etc.

10. T. *globulosa* (N. ab E.:) umbella simplici semicompositaque involucro subtetraphyllo obtuso multo longiori, spiculis solitariis ovato-subglobosis, squamis ovatis obtusis submucronatis, culmo teretiusculo aphyllo basi vaginato.—Scirpus globulosus, *Vahl. en.* 2. *p.* 277; *Willd. sp. pl.* 1. *p.* 301.—Isolepis globulosa, *R. et Sch. s. v.* 2. *p.* 119.—Fimbristylis umbellata, *Vahl. en.* 2. *p.* 291; *R. et Sch. s. v.* 2. *p.* 99.—S. umbellaris, *Lam. illust. gen.* 1. *p.* 141. ——Ceylona; *Macrae.*

In nostra caryopsis obovata, brevis tuberculis seriatis aspera, nec lævis, et stylus trifidus.

[Huic valde affinis est *Sc. pentagonus*, Roxb. (tab. pict. 746, sub *S. dichotomo*), in quo umbella subcomposita 9–12-stachya radiis spiculas subternas ovatas gerentibus, culmus aphyllus basi vaginatus, ac folia nulla.—Arn.]

§ 3. *Capitatæ aut subcapitatæ, spiculis in radiolis sessilibus aut saltem brevissime pedicellatis, 3–5, capitulum exhibentibus, teretibus, magusculis, culmo trigono aut semitereti compresso, foliis rigidis obtusis curvis aut tortis, plerisque radicalibus.*

11. T. *torta* (N. ab E.:) involucro di-triphyllo e basi membranacea lineari obtuso umbella decomposita biradiata laxiuscula multo breviore, culmo sulcato-angulato compressiusculo foliisque linearibus recurvis ex apice rotundato mucronatis concoloribus rigidulis contortis, spiculis ternis sessilibus aut brevipedicellatis oblongis, squamis obtusis glabris rufo-fuscis undique albo marginatis triandris, caryopsi obovata trigona alba tuberculata, rhizomate crasso repente fibroso fibris flagellaribus.—*Wight. cat. n.* 1885.——Peninsula Indiæ orientalis; *Wight*; *Rottler.*

Quod ad habitum accedit *Tr. junciformi*, N. ab E., a qua differt spiculis multifloris duplo et triplo majoribus involucellisque e basi latiuscula setaceis, nec squamis spicularum conformibus. Radicem si spectas et spiculas, propinquiorem existimabis *Tr. chætorrhizæ*, N. ab E. Sed nostræ culmus (semipedalis circiter) irregulariter angulatus est, nec trigonus, folia latiora sunt, involucra breviora, spiculæ majores, umbellæ radii $\frac{1}{4}$–7 lin. longi, graciliores et patuli, alter tristachyus, alter 4–5-stachyus, spiculis binis sessilibus, binis ternisve radiolo brevissimo sublatis. Caryopsis omnino ut in *Tr. chætorrhiza.*

[In exemplo altero herb. Wight. culmus plusquam pedalis, umbella fere duos pollices longa, spiculæ subquinæ omnes centrali excepta longiuscule pedicellatæ.—ARN.]

12. T. *chætorrhiza* (N. ab E.:) involucro di-triphyllo e basi membranacea subulato umbella decomposita erecta stricta multo breviori, culmo compresso-trigono stricto, foliis anguste linearibus recurvis obtusis culmo brevioribus concoloribus, spiculis subcapitatis ovatis, squamis obtuse carinatis acutiusculis lævibus fuscescentibus, caryopsi obovata trigona alba tuberculata, rhizomate crasso repente fibroso.—α; culmo distincte trigono.—*N. ab E. in herb. Royle. n.* 41.—β; culmo magis compresso.—*Wight. cat. n.* 1886.——α In Nepalia.—β In Peninsula Indiæ or.; *Wight.*

Affinis *Tr. obtusifoliæ.* Stylus est trifidus. Spiculæ in radiis simpliciter aut composite fasciculato-umbellatæ, brevissime pedicellatæ. Var. β nescio an potius cum *Tr. latifolia* sit conjungenda.

13. T. *latifolia* (N. ab E.:) involucro 2–3-phyllo subulato umbella decomposita contracta laxiuscula multo breviori, culmo sulcato-angulato superne compresso gracili, foliis lato-linearibus obtusis margine parum scabris glaucis contortis, spiculis capitatis ternis quinisve ovatis angulatis, squamis ovatis obtusis fuscis margine omni membranaceis carinatis subtilissime exasperatis, stylo basi bulboso glabro, ovario trigono.—*Wight. cat. n.* 1887.——Peninsula Ind. or.; *Wight.*

Differt a *Tr. junciformi* foliis latioribus, brevioribus, $\frac{3}{4}$–$\frac{1}{2}$-poll. longis, lineam latis, valde contortis glaucis, cæterum, ut illius folia, subtus cum vaginis suis confertim striatis, culmo et umbellæ radiis gracilioribus, spiculis angulatis, quandoque compressis plurifloris, e squamis trispiris (9–15) latioribus obscurius fuscis carina autem omnique margine latius pallidis setulis exiguis non nisi sub bona lente distinguendis appressis scabris et nonnihil micantibus, involucellis e basi ovata setaceo-aristatis.

14. T. *junciformis* (N. ab E.:) involucro di-tetraphyllo e basi latiori subulata umbella decomposita erecta stricta duplo breviore, culmo tereti-compresso sulcato, foliis linearibus recurvis obtusis culmo brevioribus concoloribus, spiculis capitatis ternis ovatis paucifloris, capitulis globosis bracteis duabus squamiformibus suffultis, squamis ovatis acutiusculis brunneis lævibus margine pallidioribus, stylo basi bulboso, superne scabro, ovario trigono glabro, culmo basi bulboso.—*N. ab E. in herb. Royle. n.* 54.—Fimbristylis, *Wall. cat. n.* 3520, *a.*——Nepalia; *Royle; Wallich.* Ad ripas Irawaddi fluminis; *Wallich.*

Simillima *Tr. chætorrhizæ*; differt autem, culmo haud trigono, sed e tereti compresso, basi bulboso, foliis latioribus, spiculis in capitula densa bibracteata ternatim compactis, 3–5-floris, squamis carina parum viridibus, cæterum ferrugineis, margine multo angustiori spatia albidis. An utræque conjungendæ?

[Exemplum meum Wall. n. 3520, *a,* vix hujus loci, vel saltem characteri supra dato haud convenit; illi umbella laxiuscula decomposita, spiculis lateralibus graciliter pedicellatis haud in capitula globosa aggregatis, et culmus compressus.—ARN.]

VIII. ISOLEPIS. *R. Br.*

§ 1. *Lateriflorae.*

1. I. *squarrosa* (Vahl.:) culmo fasciculato compresso-subquadrangulari basi unifolio vagina laxa, folio vagina breviori lineari complicato-subulato striato, spiculis 2–4-nis sessilibus ovatis, squamis e basi membranacea obo-

vata longe subulatis subula recurva scabra, stylo brevissimo trifido, caryopsi oblonga trigona lævi, involucro diphyllo e basi dilatata subulato altero capitulo longiore culmum subcontinuante.—*Vahl. en.* 2. *p.* 259 ; *R. et Sch. s. v.* 2. *p.* 111 ; *Wight. cat. n.* 1888.—Isolepis, *Wall. cat. n.* 3477.—Scirpus squarrosus, *Linn. mant. p.* 181 ; *Rottb. gram. p.* 49. *t.* 1:. *f.* 5 ; *Willd. sp. pl.* 1. *p.* 308 ; *Roxb. fl. Ind.* 1. *p.* 222 ; (*ed. Wall.*) 1. *p.* 224 ; *in cat. merc. Ind. or. mus. tab.* 751 (ex *Arn.*).—Scirpus chinensis, *Osbeck. iter, p.* 220.—*Rheed. Mal.* 12. *t.* 38 (in eadem tabula cum 36 et 37 ; tab. 39, a Willdenovio citata, *Filix* est) ; *Pluk. t.* 50. *f.* 6.—β ; spiculis majoribus, oblongo-cylindricis.—*Wight. cat. n.* 1888, *b.*—*Wall. cat. n.* 3478.——In pascuis arenosis siccis Peninsulæ Indiæ orient.; *Roxburgh ; Wight.*

Species inter omnes distinctissima squamis obovato-cuneatis aristato-cuspidatis. Haud immerito *Lipocarphæ* comparat hanc speciem cl. Vahl : at vero et ego squamulas interiores frustra quæsivi. Stylus brevissimus ab *Isolepidibus* est alienus.

2. I. *Roylei* (N. ab E. :) spicis ad speciem lateralibus aggregatis sessilibus ovatis, squamis ovatis mucronato-acutis scariosis sub frùctu erectis, folio involucrali culmum continuante erecto elongato teretiusculo articulato-nodoso, infero nullo, caryopsi obovato-trigona undique transversim undato-rugosa atra nitente, culmo subtereti glauco basi vaginato, vagina oblique truncata aphylla.—*N. ab E. in herb. Royle, n.* 48.——Nepalia ; *Royle.*

Simillima omnino *I. lupulinæ*, sed differt evidentissime parte involucrali culmi capitulum laterale superante distinctissime articulato-nodosa, interstitiis fere semipollicaribus et squamis fructuum haud patentibus.

3. I. *lupulina* (N. ab E. :) spicis ad speciem lateralibus 3–6 aggregatis sessilibus ovalibus, squamis ovatis mucronato-acutis scariosis sub fructu patentibus, folio involucrali culmum continuante erecto elongato compresso-canaliculato, infero nullo, caryopsi obovata trigona undique interrupte rugosa atra nitente, culmo tereti-compresso basi vaginato, vagina oblique truncata (in nostris) aphylla.—*N. ab. E. in herb. Royle. n.* 31.——Nepalia ; *Royle.*

I. supinæ et *commutatæ*, herb. Lindl., similis, sed distincta spiculis simpliciter aggregatis, 4–5 lin. longis, tempore maturi fructus recto angulo patentibus pallide ferrugineis scariosis. Involucri folium terminans haud raro culmum æquat, per se 1–4 poll. altum. Radix fibrosa.

4. I. *setacea* (R. Br. :) culmis fasciculatis teretiusculis setaceis basi unifoliis, spiculis 1–2 ovatis involucri subdiphylli altero foliolo brevioribus, squamis membranaceis ovatis articulatis superne venulosis 2-andris, caryopsi obovata convexa longitudinaliter (7–11-) costata interstitiis transversim striatis, stylo profunde trifido, radice fibrosa.—*Br. prodr. fl. nov. Holl.* 1. *p.* 222.—Scirpus setaceus, *Linn. sp. pl.* 1. *p.* 33 ; *Smith. fl. Brit.* 1. *p.* 54 ; *Engl. bot. t.* 1693 ; *Schrad. fl. Germ.* 1. *p.* 137 ; *Host. gram. Austr.* 3. *p.* 44. *t.* 65.—β, *fuscula* ; culmis strictiusculis 1–1½ poll. longis, squamis saturate castaneo-fuscis, carina viridi, caryopsi 7-costata.—*N. ab E. in Cyp. Royl. n.* 63.——β. In Nepalia ; *Royle.*

Stylus profunde trifidus (nec bifidus, ut habet Linkius), scilicet : bifidus, altero ramo bipartito.

5. I. *supina* (R. Br. :) culmis fasciculatis obtuse trigonis validiusculis (¼–1-pedalibus) basi unifoliis foliolo brevi lineari-canaliculato, spiculis ad speciem lateralibus simpliciter compositeve glomeratis ovato-oblongis pallidis, squamis alte carinatis ovatis acuminatis, caryopsi obovato-trigona transversim undulato-rugosa nigra pallidave, involucro diphyllo, folio altero culmum continuante erecto longissimo trigono-complicato, altero inferiore spiculis breviore.—α ; fructu nigro, foliis distinctis.—*Wight cat. n.* 1889.—Isolepis supina, *Br. prod. fl. nov. Holl.* 1. *p.* 221 ; *Vahl. en.* 2. *p.* 257 ; *Link. h. Ber.* 1. *p.* 284.—Scirpus supinus, *Linn.* ; *Willd. sp. pl.* 1. *p.* 299 ; *Schrad. fl. Germ.* 1. *p.* 38. *t.* 1. *f.* 1 ; *Host. gr. Austr.* 3. *t.* 64 ; *Roxb. fl. Ind.* 1. *p.* 217 ; (*ed. Wall.*) 1.

p. 219; *in cat. merc. I. o. mus. tab.* 1933. (fide *Arn.*)—S. lateralis, *Retz. obs.* 4. *p.* 12; *et* 5. *p.* 16; *Wall. cat. n.* 3469, *a.*—β; fructu nigro, vaginis ob laminam brevissimam subaphyllis.—*Wight. cat. n.* 1889, *b*; *Wall. cat. n.* 3461, *et* 3468 (ex parte).—γ, *leucosperma*; caryopsi alba.—*N. ab E. in herb. Royle. n.* 58. *t.* 71.——Peninsula Ind. or.; *Klein*; *Heyne*; *Wight.* Benghalia inferiori; *Roxburgh*; *Wallich.* Silhet; *Wallich.*—γ. In Nepalia; *Royle.*

6. I. *corymbosa* (Roth.:) culmo compresso (?) vagina baseos aphylla, spiculis ad speciem lateralibus corymbosis ovatis parvis, squamis ovatis scariosis mucronulatis flavo-fuscoque variis carina viridi, involucro communi monophyllo culmum continuante erecto teretiusculo corymbo paulo breviori, partialibus scariosis.—Scirpus corymbosus, *Heyne in Roth. nov. sp. pl. p.* 28.——Ex India orientali misit *Heyne.*

7. I. *prolongata* (N. ab E.:) spiculis ovatis ovatove oblongis teretibus capitatis, capitulo globoso sessili ad speciem laterali, involucro culmo triplo quadruplove longiore tereti obscurius articulato, squamis late ovatis obtusis membranaceis rufo-testaceis carina obtusa late virente trinervi, caryopsi obovata triquetra transversim rugulosa opaca fusca, culmo tereti aphyllo fistuloso.—*Wight. cat. n.* 1890.—I. articulata, *N. ab E. in Cyp. Royle. n.* 151.—Scirpus prolongatus, *Poir. enc. meth.* 6. *p.* 764; *suppl.* 5. *p.* 91; *R. et Sch. s. v.* 2. *p.* 131.—S. subarticulatus, *Roxb. fl. Ind.* 1. *p.* 215; (*ed. Wall.*) 1. *p.* 217; *in cat. merc. I. o. mus. tab.* 740 (fide *Arn.*)—*Wall. cat. n.* 3457.——In Peninsulæ Ind. or.; *Roxburgh*; *Wight.* In Nepalia, Silhet, Sagaen ad ripam Irawaddi fluvii, Kamaon; *Wallich.*

Differentia patet ex iis, quæ diximus; synonymia haud ita. Primo intuitu distinguitur hæc nostra squamis spiculæ latioribus, vix evidenter carinatis, dorso medio lanceolato spatio viridi trinervi herbacea partem membranaceam apicis haud penetrante, lateribus rufo-testaceis, margine omni pallescente. Stylus apice trifidus, ramis brevibus. Stamina tria. Caryopsis obovato-trigona, transversim undato-rugulosa, fusca (nec alba ut habet Roxb.), opaca.

8. I. *articulata* (N. ab E.:) spiculis ovato-pyramidalibus acutis capitatis, capitulo globoso ad speciem laterali, involucro culmo duplo longiore tereti intense articulato, squamis ovatis acutis membranaceis testaceis concoloribus carina angusta viridi, caryopsi obovata triquetra lævi atra nitente, culmo tereti aphyllo striato.—*Wight. cat. n.* 1891.—Scirpus articulatus, *Linn. sp. pl.* 2. *p.* 70; *Willd. sp. pl.* 1. *p.* 290; *Rottb. gram. p.* 45; *Vahl. symb.* 1. *p.* 8; *Roxb. fl. Ind.* 1. *p.* 214; (*ed Wall.*) 1. *p.* 217; *in cat. merc. I. o. mus. tab.* 739 (fide *Arn.*)—S. fistulosus, *Forsk. descr. p.* 14.—*Wall. cat. n.* 3459.—*Rheed. Mal.* 1. *t.* 71; *Moris. hist. pl.* 3. *p.* 31. *sect.* 8. *n.* 3.——In aquis stagnantibus Peninsulæ Indiæ orientalis.

Scirpum fistulosum, Forsk. diversum esse ab *articulato* docet Link. (Jahrb. 3. p. 79). *S. articulatus*, Vahl. excludendus, quippe cui perigynium setosum.

9. I. *incurvata* (N. ab E.:) spiculis ovato-oblongis capitatis, capitulo ad speciem laterali late ambiente culmi basi propinquo, involucro culmo multo longiore tereti intense articulato, squamis ovatis obtusis, caryopsi obovata trigona.—Scirpus incurvatus, *Roxb. fl. Ind.* 1. *p.* 214; (*ed. Wall.*) 1. *p.* 216; *in cat. merc. Ind. or. mus. tab.* 1932 (fide *Arn.*)—S. articulatus β tenuis, *Roth. nov. pl. sp. p.* 26.——Locis humidis circa Calcuttam; *Roxburgh.*

Differt ab *I. articulata* et *prolongata* culmis tenerioribus, 6–9 pollices longis, ita incurvis, ut cæspes globi figuram adipiscatur.—Deest in collectionibus a me visis.

§ 2. *Terminales.*

10. I. *trifida* (N. ab E.:) spicis terminalibus subternis oblongis, lateralibus longe pedicellatis patentibus, squamis ovatis carinatis acutis, involucro subdiphyllo foliolis e basi ovata membranacea setaceis spica sessili brevioribus, caryopsi obovata trigona transversim crenulato-subrugosa nigro-mucronata,

culmo filiformi basi unifolio, folio setaceo canaliculata culmo multo breviori, vagina ore barbata.—*N. ab E. in herb. Royle. n.* 51.——Nepalia; *Royle.*

Ab. *I. brevifolia*, Bernh., (*Scirpo capillari*, Muhl., seu *Muhlenbergiano*, Spr.) differt spiculis fuscis, lateralium pedicellis spicula sua plerumque longioribus.

11. I. *densa* (Wall.:) spiculis ovalibus angulatis in umbellam simplicem aut compositam dispositis, radiolis (radiisve) unifloris elongatis, squamis trifariis ovatis acutis muticis alte carinatis fuscis glabris monandris, caryopsi obovata trigona confertim et irregulariter transverse granulato-rugulosa lutescente, involucri subdiphylli foliolis e basi scariosa setaceis umbella brevioribus, involucellis submuticis, culmo filiformi sulcato-angulato glabro, vaginis apice barbatis, foliis setaceis (canaliculatis) brevibus.—*Wall. in Roxb. fl. Ind.* (*ed. Wall.*) 1. *p.* 231.—Fimbristylis, *Wall. cat. n.* 3514, *c, d, e.*—Isolepis, *Wall. cat. n.* 3476.——In Nepalia, Kamaon, Benghalia inferiori, et montibus Silhet; *Wallich.*

Similis *I. gracili* et *barbatæ*, sed differt præter characteres supra relatos, elegantia quadam structura et umbellæ gracilitate atque liberiori evolutione.

12. I. *gracilis* (N. ab E.:) spiculis oblongis angulatis capitato-geminis ternisve, capitulis in umbellam compositam decompositamve depresso-patentem collectis aut simpliciter aggregatis, squamis trifariis ovatis alte carinatis mucronato-acutis rufo-fuscis scabris monandris, caryopsi obovata trigona transversim rugosa lutescente, involucri et involucelli foliolis e basi ovata scariosa subulato-cuspidatis umbella umbellulaque brevioribus longioribusve, culmo filiformi tetragono compresso, vaginis apice dehiscentibus villosis, foliis setaceis.—*α*; spiculis pluribus umbellatis, foliis brevibus.—*Wight. cat. n.* 1892.—*β*; spiculis subternis vel pluribus lateralibus pedicellatis, involucro diphyllo multo brevioribus, culmo capillari, foliis dimidium culmum attingentibus.—*Wight. cat. n.* 1892, *b.*——Peninsula Ind. orient.; *Wight.*

Ab *Isolepide barbata* differt spiculis scabris in capitula umbellaria dispositis nec aggregatis, et fructu transversim rugoso; ab *Isolepide arenaria* umbellari inflorescentia et squamis alte carinatis trifariis.

13. I. *barbata* (R. Br.:) capitulo subgloboso, spiculis multis capitato-aggregatis compresso-trigonis, squamis ovatis carinatis infra apicem mucronatis, involucro diphyllo capitulo breviori squamis simili, foliis setaceis, vaginis arctis ore barbatis, caryopsi obovato-trigona lævi.—*N. ab E. in Linnæa*, 8. 1, *p.* 83; *R. et Sch. s. v.* 2. *p.* 112; *Wight. cat. n.* 1893.—Scirpus barbatus, *Rottb. gr. p.* 52. *t.* 17. *f.* 4; *Roxb. in cœt. merc. I. o. mus. tab.* 753 (ex *Arn.*)—S antarcticus, *Vahl. en.* 2. *p.* 261 (excl. synon. *Linn.*); *Roxb. fl. Ind.* 1. *p.* 223; (*ed. Wall.*) 1. *p.* 225.—S. capillaris, *Linn. sp. pl.* 1. *p.* 73; *fl. Zeyl. p.* 39; *fl. Cliff.* 21.—*Wall. cat. n.* 3481.—*α*; culmo longiori, $\frac{1}{3}$-$\frac{3}{4}$ ped. alto, capitulo pleiostachyo.—(1; vaginis sanguineis.—2; vaginis pallidis.—I. fascicularis, *Sieb. fl. Seneg.*—*Wall. cat. n.* 3481, *d, e*).—*β*, *Roxburghiana*; culmo humiliori stricto, capitulo 3–5-stachyo.—Isolepis, *Wall. cat. n.* 3497.—[Scirpus monander, *Roxb. fl. Ind.* 1. *p.* 222; (*ed. Wall.*) 1. *p.* 225.—S. pygmæus, *Roxb. in cœt. merc. I. o. mus. tab.* 752.—Arn.]—*γ*, *capillaris*; culmo capillari pollicari, spiculis capitato-ternis pallidis involucro diphyllo brevioribus, vaginis pallidis.—*N. ab E. in herb. Royle. n.* 50.—*Burm. herb. Zeyl. t.* 47. *f.* 2.—*δ*, *recurvata*; culmo 1–2-pollicari foliisque setaceis recurvis; his culmum æquantibus scaberrimis, spiculis capitato-ternis quinisve coloratis involucro di-triphyllo brevioribus, vaginis pallidis.—*Wight. cat. n.* 1893, *b.*——*α*. Ceylona; *Macrae.* China; *Vachell.* Peninsula Ind. or.; *Roxburgh*; *Wight*; &c. Nepala, Melloon, Taong-Dong, Ava, et ad ripam Irawaddi; *Wallich.* Gongachora; *Hamilton.*—*β*. Promiscue cum *α*. Senegal; *Sieber.*—*γ*. Nepala; *Royle.* Ceylona; *Burmann.*—*δ*. Peninsula Ind. or.

XIX. ELEOGITON. *Link.*

1. E. *curvulus* (N. ab E.:) culmo ramoso folioso flaccido repente natanteve, foliis fasciculatis linearibus acutiusculis trinervibus incurvis culmum tereti-compressum striatum incurvum subæquantibus, spica solitaria terminali, squamis subrotundis truncatis inferioribus spica brevioribus, staminibus duobus, caryopsi obovata lenticulari lævi.—*N. ab E. in Cyp. Lindl.; Wight. cat. n.* 1894.——In Ceylona; *Macrae.* In Peninsula Indiæ orientalis; *Wight.*

Ab *E. fluitante,* cui maxime similis, differt: foliis latioribus brevioribus rigidioribus culmos (e fasciculis, ut in illa, nascentes) et latitudine et longitudine adæquantibus, nec iisdem angustioribus, spicæ squamis subæqualibus, plerumque ad latera purpura tinctis.

XX. MALACOCHÆTE. *N. et Meyen.*

1. M. *pectinata* (N. ab E.:) culmo tereti apice trigono, panicula composita, laminis hypogynis cuneiformibus pinnatifido-ciliatis.—*N. ab E. in herb. Royle. p.* 56; *in Ed. n. phil. journ.* (1834) *n.* 34. *p.* 264; *Wight. cat. n.* 1895.—Scirpus pectinatus, *Roxb. fl. Ind.* 1. *p.* 218; (*ed. Wall.*) 1. *p.* 220.—S. campestroides, *Roxb. in cœt. merc. I. o. mus. tab.* 744 (fide *Arn.*)—S. plumosus, *R. Br.? prod. fl. nov. Holl.* 1. *p.* 223.——Locis turfosis arenosis Peninsulæ Indiæ orientalis; *Roxburgh; Wight.* Nepalia; *Royle.*

XXI. HYMENOCHÆTE. *P. de B.*

1. H. *grossa* (N. ab E.:) spiculis decomposito-corymbosis, involucro subtriphyllo, culmi angulis lævibus.—*N. ab E. in Ed. n. phil. journ.* (1834) *p.* 264; *Wight. cat. n.* 1896.—Isolepis, *Wall. cat. n.* 3470.—Scirpus grossus, *Retz. obs.* 5. *p.* 15; *R. et Sch. s. v.* 2. *p.* 141; *Vahl. en.* 2. *p.* 270.—[*Roxb. fl. Ind.* 1. *p.* 231; (*ed. Wall.*) 1. *p.* 230.—I. giganteus, *Roxb. in cœt. merc. I. o. mus. tab.* 764.—ARN.]——In Peninsulæ Ind. or. aquis dulcibus stagnantibus profundioribus; *Roxburgh; Wight.* Nathpoor et Gongachora; *Hamilton.*

Species, notis in generis charactere l. c. indicatis sat distincta. *Scirpus maximus,* Roxb. (in cœt. merc. I. o. mus. tab. 765 (fide *Arn.*) forsan eadem planta est, setis jam delapsis observata. Culmus crassus, argute triquetes, angulis lævibus. *Scirpus Kysoor,* Roxb. (in cœt. merc. I. o. mus. tab. 2017 (fide *Arn.*) angulis culmi retrorsum scabris, et setis in fructu persistentibus differre videtur.

[Roxburghius nullibi monuit setas in *S. Kysoor* persistere, et eas eodem fere modo discripsit ut in *S.* suo *grosso.*]

XXII. ERIOPHORUM. *Linn.*

Spiculæ undique imbricatæ squamis membranaceis. Perigynium constans e setis plurimis capillaribus basi fasciculatim cohærentibus squamas plerumque longe superantibus lanuginem aut comam exhibentibus. Stylus trifidus, longus. Caryopsis trigona, mucronata.—Inflorescentia: Spiculæ raro in apice culmi solitariæ, plerumque in cymam magis minusve compositam involucratam distributæ.

1. E. *comosum* (Wall.:) culmo subtrigono, foliis lineari-filiformibus canaliculatis margine scabris, anthela supradecomposita involucrata, perigynii coma squama triplo quadruplove longiore.—*Wall. cat. n.* 3446.—*N. ab E. in herb. Royle. n.* 79, 80.—Scirpus comosus, *Wall. in Roxb. fl. Ind.* (*ed. Wall.*) 1. *p.* 234.—β, *nanum;* culmo vix bipollicari distachyo, foliis perangustis.—*N. ab E. in herb. Royle. n.* 78.——Nepalia; *Wallich; Royle.*

2. E. *arundinaceum* (Wall.:) culmo obtuse subtrigono alto, foliis linearibus acute canaliculatis margine scaberrimis rigidis, anthela supradecomposita

coarctata involucrata radiis elongatis, perigynii coma squama duplo triplove longiore.—*Wall. cat. n.* 3448, *b.*——E. Penang; *Wallich.*

Quod ad formam anthelæ et comam mediocrem sordide albam attinet, simile *E. comoso*, sed differt culmo altiore, 2–3-pedali, foliisque latioribus.

XXIII. SCIRPUS. *Linn.*; *Lestib.*

In hoc genere stylum aut bifidum aut trifidum cum caryopsi depressa magis minusve compressave conjunctum invenimus, quo sequitur, divisionem ejus, quod ad fructum, accessoriam hic esse, nec essentialem; id quod id ex eo patere videtur, quod stylus primo loco bifidus evidenter inveniatur, tum ramus alter paulo superius denuo divisus, trifidum exhibent stylum, ubi vero hæc ulterior divisio supprimatur, bifidus remaneat.—Stylus trifidus cum caryopsi trigona conjunctus, characterem præbet haud spernendum, licet non in omnibus magni momenti sit; idem de stylo bifido cum caryopsi depressa compressave (exangulata). Ubi caryopsis depressa compressave tam cum stylo trifido quam cum bifido jungitur, styli subdivisio haud respicienda. Idem etiam de caryopsi trigona cum stylo bifido conjuncta censeres, si talis casus sano statu occurrerit.

§ 1. *Culmo folioso, umbella terminali.*

1. S. *maritimus* (Linn.:) culmo trigono, basi folioso, foliis linearibus carinatis elongatis, spiculis terminalibus ovatis oblongisve in umbellam simplicem compositamve dispositis tumque in radiis subglomeratis totave umbella glomeratim contracta, squamis ferrugineis lacero-tridentatis e carina aristato-mucronatis, involucro umbellam superante, caryopsi ancipiti lævi.—*Linn.*; *Link. h. Ber.* 1. *p.* 298.—*ι*, *glaucus.*—*ι** : radiis di-tristachyis.—*R. et. Sch. s. v.* 2. *p.* 149; *N. ab E. in herb. Royle. n.* 99.—S. glaucus, *Lam. enc. meth.* 1. *p.* 142; *Poir. suppl.* 6. *p.* 769.—S. tridentatus, *Roxb. fl Ind.* 1. *p.* 225; (*ed. Wall.*) 1. *p.* 228; *in cat. merc. I. o. mus. tab.* 755 (fide *Arn.*)—*Wall. cat. n.* 3504.—*ι***, *cylindricus*; radiis monostachyis, spiculis oblongis in umbella quadriradiata, binisque centralibus sessilibus, involucro subtriphyllo, foliolo uno umbella longiore.—*Wight. cat. n.* 1897.——Peninsula Indiæ orientalis; *Roxburgh*; *Wight.*

2. S. *affinis* (Roth.:) culmo trigono inferne folioso foliisque linearibus lævibus, spiculis terminalibus paucis grandiusculis capitatis ovato-oblongis, squamis luteis margine pallidioribus apice lacero-dentatis nervoque carinali squarroso-aristulatis, inferioribus trinervibus, caryopsi obovata compressa lævi in fundo grisco subtilissime fusco-punctulata, stylo bifido, involucro diphyllo capitulo multo longiori.—*Roth. nov. pl. sp. p.* 30; *R. et Sch. s. v.* 2. *p.* 140; *N. ab E. in Royle. Cyp. n.* 45, 65, *et* 66.—S. cephalotes, *Heyn.*—*Wall. cat. n.* 3463.——Peninsula Indiæ orientalis; *Heyne.* Monghir; *Hamilton.* Agsufghur; *Wallich.* Nepalia; *Royle.*

Spiculis majoribus pallidioribus semper capitatis et fructu compresso facile distinguitur a *S. maritimo.* Haud improbabile est hunc esse verum *S. cephalotem*, Linn.

[In *S. maritimo* Europæano spicas sæpe paucas sessiles et capitatas offendo, nec limites immutabiles inter hasce duas species indicare possum.—ARN.]

§ 2. *Culmo aphyllo, inflorescentia ad speciem laterali.*

3. S. *triangulatus* (Roxb.:) culmo a basi acute triquetro, capitulo spurie laterali sessili denso involucro terminante acute triquetro recto duplo breviore, spiculis ovatis, squamis subrotundis mucronatis striatis brunneo-fuscis nitentibus, stylo trifido, caryopsi plano-convexa obovata lævi setas rigidas æquante.—*Roxb. fl. Ind.* 1. *p.* 217; (*ed. Wall.*) 1. *p.* 229; *Wall. cat. n.* 3467.——In Chittagong; *Roxburgh.* Nepalia et Sillet; *Wallich.* Monghir, Amosha, et Sanashygota; *Hamilton.* Ceylona; *Macrae.*

Culmo humiliori, argutissime triquetro et capitulo brunneo nitente parum ab apice distante a sequente specie distinguitur. An ambæ varietates unius speciei?

4. S. *Javanus* (N. ab E. :) culmo inferius subtrigono apice triquetro, capitulo polystachyo sessili denso involucro terminante recto acute triquetro triplo breviore, spiculis oblongis, squamis ovatis mucronatis striatis virescentibus fusco-marginatis, stylo trifido, caryopsi obovata plano-convexa lævi setas rigidas æquante.—*N. ab E. in Royle Cyp. n.* 52.——In Java; *Blume.* Ceylona; *Macrae.* Nepalia; *Royle.*

Differt a præcedente statura paulo majore, culmo basin versus compresso-trigono, acumine longiore, spiculis majoribus, adeo in capitulo viridulis opacis.

5. S. (Heleophila?) *Wallichii* (N. ab E. :) culmo aphyllo involucroque erecto monophyllo longo tetragonis, vagina baseos truncata, spiculis subgeminis ad speciem lateralibus ovatis acutis, squamis ovatis acutis margine albo-membranaceis in disco viridibus viridi-lineolatis, caryopsi obovato-subrotunda altero latere convexa altero concava transversim undato-subrugosa, stylo bifido.—*Wall. cat. n.* 3468 (ex partè).——Silhet; *Wallich.*

Ob similitudinem *S. junciformis* inter *Scirpos* reliquimus, etiamsi (in uno specimine) omnes, quos vidi, styli bifidi (et completos equidem vidi) observarentur.

6. S. *junciformis* (N. ab E. :) culmo tereti, vaginis baseos aphyllis, florali culmo conformi eundemque continuante capitulum multoties superante incurvo, spiculis ad speciem lateralibus 1–5 arcte sessilibus ovato-conicis, squamis subrotundis mucronulatis dorso herbaceo viridibus uninervibus ambitu membranaceis pallidis undique fusco-lineolatis glabris, stylo bifido altero ramo bifido, caryopsi obovata plano-convexa punctulata subtilissimeque oblique rugulosa pallida, setis rigidis denticulis retro spectantibus hispidis caryopsin æquantibus.—*Wight. cat. n.* 1898.—S. juncoides, *Roxb. fl. Ind.* 1. *p.* 226; (*ed. Wall.* 1. *p.* 218.—S. supinus, *Roxb. in cœt. merc. I. o. mus. tab.* 742 (fide *Arn.*) —*Wall. cat. n.* 3469 excl. *a.*)——Peninsula Indiæ orientalis locis sidis uliginosis. In Benghalia inferiori, in Nepalia, Silhet; *Wallich.* In Java; *Blume.*

Stylus minime bifidus, sed trifidus, scil. altero ramo paulo minus profunde diviso pro bifido, nisi accuratius circumspicias, ad speciem tantum bifidus. Squamis rigidulis dorso medio viridulis a plerisque confinium distinguitur, tum vero et habitu *Junci filiformis* grandefacti; a *S. Wallichii* nostro culmo distinguitur tereti.

[Omnino eadem species, ut mihi videtur, est *Sc. debilis*, Torrey et Bigelow (exemplis a Torrey et Greene communicatis), forsitanque etiam Muhlenbergii quanquam descriptio in paucis haud quadrat.—ARN.]

7. S. *Luzonensis* (Presl.) culmo tereti, vagina mutica vel brevissime aciculata, involucro monophyllo longissimo culmum continuante, spiculis lateralibus subternis sessilibus ovatis, squamis late ovatis mucronulatis fusco-lineolatis nervo unico viridi, stylo trifido, caryopsi obovata plano-convexa rugulosa, setis perianthialibus sex caryopsin acquantibus.—*Presl. rel. Hœnk.* 3. *p.* 193.——In Manila, ad lagunam de Bay; *Meyen.* Ad Portum Sorzogon; *Hœnke.*

Descriptioni cl. Preslii addimus: spiculæ sæpe et geminatæ; squamæ dorso viridulæ, latere pallidæ, lineolis ferrugineis irroratæ. Vaginæ culmorum truncatæ, foliolo exili subulato vix linea longiori culmoque adpresso præditæ.

XXIV. ELEOGENUS.—Eleophila. *Pal. de Beauv.*—Heleogiton. *Schult.*

Eleogiton cognomen jam a cl. Linkio aliter usurpatum; novum igitur *Eleogeni* nomen, cum *Heleophila* nimis facile cum *Heliophila* commutari possit, imposui.

§ 1. *Genuini, perianthio præditi.*

1. E. *capitatus* (N. ab E. :) radice annua, culmo breviusculo, spica ovato-subrotunda, squamis obtusis, perigynio setoso.—*Wight. cat. n.* 1899.—Eleo-

charis capitata, *R. Br. prod. fl. nov. Holl.* 1. *p.* 225 ; *R. et Sch. s. v.* 2. *p.* 153. —Scirpus capitatus, *Willd. sp. pl.* 1. *p.* 294 ; *Vahl. en.* 2. *p.* 250 ; *Humb. et K. nov. gen. et sp. pl.* 1. *p.* 181 ; *Roxb. fl. Ind.* 1. *p.* 216 ; (*ed. Wall.*) 1. *p.* 218 ; *in cat. merc. Ind. or. mus. tab.* 741 (fide *Arn.*)—S. Caribæus, *Rottb. gram. p.* 46. *t.* 15. *f.* 3.—Isolepis? *Wall. cat. n.* 3486. *et* 3493.—α ; major, spica sphærica, squamis pallidis.—*Wall. cat. n.* 3493 (ex *herb. Wight.*).—β ; minor, spica ovato-globosa, squamis fuscis margine pallidis.—γ ; idem, squamis pallidis.—*Wall. cat. n.* 3486. *c.*—δ ; minor, culmis recurvis, spicæ ovato-pyramidalis squamis pallidis.——Bahiæ in humidis. Maranhao ; *Don.*—α, γ, et δ, Peninsula Indiæ orientalis ; *Roxburgh* ; *Wight*. In Bengalia inferiore ; *Wallich*. —β In Brasilia ad Rio de Janeiro. In Ins. St Croix.

§ 2. *Heterocliti, perigynio destituti.*

2. E. *atropurpureus* (N. ab E. :) monandrus, spicula solitaria ovata squamisque ovatis margine purpureis obtusis, caryopsi obovata lævi atra basi constricta stipite candido, culmo tereti aphyllo, vagina atropurpurea.—*Wight. cat. n.* 1900.—Scirpus atropurpureus, *Retz. obs.* 5. *p.* 14 ; *Willd. sp. pl.* 1. 1. *p.* 294.—Isolepis atropurpurea, *Vahl. en.* 2. *p.* 249 ; *R. et Sch. s. v.* 2. *p.* 106. ——In pascuis humidis Indiæ orientalis ; *Koenig.* ; *Roxburgh* ; *Wight.*

Quod ad fructum et universum habitum *Eleogeno* observante jam Retzio persimilis, nisi quod careat perigynio, cujus quidem vestigia Retzius vidit, nos autem frustra quæsivimus ; nec Vahlius ejusmodi aliquid reperit. Ex charactere igitur styli et fructus, perigynii carentis *Dichostylis* est Palissotiana ; nos autem hac in re affinitati luculentissimæ consulere maluimus. Roxburghii *Scirpus* hujus nominis dubius est, cum foliis præditus esse dicatur.

XXV. ELEOCHARIS. *R. Br.*

A *Limnochloa* differt *Eleocharis* squamarum textura tenuiori et scariosa.

§ 1. *Stylo bifido.*

1. E. *tetraquetra* (N. ab. E. :) culmo tetraquetro tenui, spica oblonga densa, squamis ovalibus obtusis rigidulis vitta intramarginali fusca, caryopsi obovata lævi, rostro caryopsin æquante, setis rufis altitudine caryopseos cum ejus rostro.—Limnochloa tetraquetra, *N. ab E. in cyp. Royl. n.* 57.—*Wall. cat. n.* 3452.——Nepalia ; *Wallich* ; *Royle.*

2. E. *palustris* (R. Br. :) culmo basi stolonifero tereti striato enodi vaginis truncatis, spica ovali aut cylindrica basi squamis duabus suboppositis subrotundis, reliquis oblongis membranaceis fusco-bivittatis, caryopsi ancipiti lenticulari fulva, styli basi bipartita corrugata.—*Link. h. Ber.* 1. *p.* 280.—β, *australis* ; squamis spicæ acutiusculis, totis albis tenuibus, nervo medio pallide viridi litura cuspidata fusca cincto.—*N. ab E. in herb. Royl. n.* 97.—*Wall. cat. n.* 3451.—ι, *humilis* ; spica ovali, squamis obscurioribus albo marginatis, vagina oblique truncata angulo acuto pallida fusco punctata.—*N. ab E. in Royle. herb. n.* 150.——β, In Owahu ins. Sandvicensi.—α et ι, Nepalia ; *Wallich* ; *Royle.*

An hæc (β et ι) species distinctæ ? At vero nullis, nisi quos supra adnotavimus, characteribus gaudent.

3. E. *uniglumis* (Link. :) culmo basi stolonifero tereti striato enodi, vaginis truncatis, spicula ovali basi squama singula semicirculari basi viridi lato ambitu fusca totam fere spiculam ambiente, reliquis squamis ovatis obtusiusculis fusca totam fere spiculam ambiente, reliquis squamis ovatis obtusiusculis fuscis margine albis carina angusta virente, styli bipartiti basi crassissima æquali virente.—*Link. h. Ber.* 1. *p.* 281 ; *in jahrb. d. Gewachsk.* 3. *p.* 77 ; *Sch. mant. s. v.* 2. *p.* 83 ; *Mertens et Koch Deutschl. fl.* 1. *p.* 426 ; *N. ab E. in cyp. Royl. n.* 62.——In Germania et America septentrionali. In Nepalia superiori ; *Royle.*

XXVI. LIMNOCHLOA. *P. de B.*

1. L. *plantaginea* (N. ab E.:) culmis teretibus dissitiuscule interceptis striatis, spica subcylindrica, squamis late ovalibus obtusis anguste marginatis rigidulis lævibus nervo medio carinali, caryopsi .....—*Wight. cat. n.* 1901.—Scirpus plantagineus, *Rottb. gram. p.* 45. *t.* 15. *f.* 2; *Vahl. en.* 2. *p.* 251; *Roxb. Ind.* 1. *p.* 212; *(ed. Wall.)* 1. *p.* 214.—S. interceptus, *Roxb. in cæt. merc. Ind. or. mus. tab.* 734 (fide *Arn.*)—Eleocharis plantaginea, *R. et Sch. s. v.* 2. *p.* 150.—*Wall. cat. n.* 3454.——In Peninsulæ Indiæ orientalis aquis stagnantibus; *Roxburgh*, &c. Bengalia inferiore; *Wallich.* Ceylona; *Macrae.*

Scirpus tuberosus, Roxb. fl. Ind. 1. p. 210, (ed. Wall.) 1. p. 213, et in cœt. merc. I. o. mus. tab. 1112 (fide *Arn.*), nil nisi hujus speciei forma bulbifera esse videtur.

[Wall. cat. n. 3512 valde differt, quamvis huc citet "*Sc. plantagineum*, Roxb.," minime fl. Ind. At suspicor, loco: *Sc. plantagineus*, herb. Roxb., legendum esse in Wall. cat.: *Sc. pentagonus*, herb. Roxb., et tab. pict. n. 762, seu *Sc. quinquangularis*, Roxb. fl. Ind. *Scirpus tumidus*, Roxb., huic valde affinis; an varietas?—ARN.]

2. L. *tumida* (N. ab E.:) culmis teretibus confertim interceptis septis tumidulis, spica subcylindrica, squamis ovato-lanceolatis acutis nervo medio carinali, caryopsi .....—Scirpus tumidus, *Roxb. fl. Ind* 1. *p.* 212; (*ed. Wall.*) 1. *p.* 215; *in cœt. merc. I. o. mus. tab.* 735 (fide *Arn.*)——In Peninsula Indiæ orientalis; *Roxburgh.*

3. L. *media* (N. ab E.:) culmo trigono compresso striato, vaginis oblique truncatis lobulo acuminato, squamis late ovalibus obtusis disco lato striatis ambitu membranaceis sublobatis pallidis, fructu ... —Scirpus medius, *Roxb. fl. Ind.* 1. *p.* 213; (*ed. Wall.*) 1. *p.* 216.—S. submutatus, *Roxb. in cœt. merc. I. o. mus. tab.* 738 (fide *Arn.*)——In Peninsula Ind. or. ad montes Circarenses; *Roxburgh.* Ceylona; *Macrae.*

An *L. spirilis* var.? ; differt squamis concoloribus disco lato ovali striatis, et culmo irregulariter trigono compressiusculo.

4. L. *acutangula* (N. ab E.:) culmo acute triquetro, vaginis oblique truncatis, spicæ cylindricæ acutæ spiris paucis diclinibus, squamis late ovalibus obtusis disco herbaceo lato striato viridulo ambitu membranaceis sublobatis pallidis, caryopsi obovata costata interstitiis transversim striatis.—*Wight. cat. n.* 1902.—Scirpus acutangulus, *Roxb. fl. Ind.* 1. *p.* 213; (*ed. Wall.*) 1. *p.* 216. —S. mutatiformis, *Roxb. in cœt. merc. I. o. mus. tab.* 737 (fide *Arn.*)—Eleocharis, *Wall. cat. n.* 3453.——In aquis quietis stagnantibns Peninsulæ Indiæ orientalis. Benghala; *Hamilton*; *Wallich.*

A *L. spirali* maxime differt spiris squamarum evidentissimis, et squamis aliter contractis. Squamæ, ex elliptico subrotundæ et obtusæ, quandoque apicem versus a latere inflectuntur eoque nonnihil attenuatæ ("somewhat acute") nec vero acutæ dicendæ. Caryopsis pallida. Perigynii setæ sex, planiusculæ, rigidæ, apice denticulis retroversis præditæ, caryopsin æquantes.

5. L. *spiralis* (P. de B.:) culmo triquetro, vaginis oblique truncatis, spicæ oblongo-cylindricæ obtusæ spiris pluribus obliquis, squamis late cuneiformibus truncato-subtrilobis obtuse carinatis, caryopsi trigona.—Scirpus spiralis, *Rottb. gram. p.* 45. *t.* 15. *f.* 1; *Willd. sp. pl.* 1. *p.* 290; *R. et Sch. s. v.* 2. p. 155; *Roxb. fl. Ind.* I. *p.* 212; (*ed. Wall.*) 1. *p.* 215.—S. mutatus, *Roxb. in cœt. merc. I. o. mus. tab.* 736 (fide *Arn.*)——In aquis stagnantibus quietis Peninsulæ Ind. or.; *Roxburgh.*

Inter plantas tam Wallichianas quam Wightianas hæc species desideratur.

TRIB. IV. RHYNCHOSPOREÆ.

XXVII. RHYNCHOSPORA. *Vahl.*

Spiculæ uni-aut sesquifloræ, squamis subtrifariis, inferioribus minoribus, sterilibus, penultima fructifera. Perigynii setæ sex, denticulis antrorsum spectantibus scabræ, caryopsin cum ejus rostro æquantes. Stylus ovario articulo conjunctus, profunde bifidus. Caryopsis obovata, compressiuscula, rostro lato discreto terminata.—Inflorescentia: spiculæ fasciculatæ, fasciculis lateralibus et terminali magis composito corymbosis. Vaginæ truncatæ ore membranaceo.

1. R. *Chinensis* (N. et Meyen:) culmo triquetro angulis scabro, corymbis axillaribus alternis compositis pedunculatis inferioribus distantibus, spiculis subgeminis, rostello conico longitudine caryopseos obovatæ transversim scrobiculato-rugosæ, foliis linearibus canaliculatis margine pedunculisque scaberrimis.—Rhynchospora, *Wall. cat. n.* 3421.——In China; *Meyen.* In Nepalia; *Wallich.*
A *Rh. laxa* Vahlii differt foliis angustioribus et corymbulis distantibus. A *Rh. Brunei* et *castanea* culmo triquetro. A *Rh. triflora* Vahlii vix differt nisi corymbulis axillaribus plurifloris, et vereor ne hæc sit mera *Rh. Chinensis* varietas.

XXVIII. MORISIA. *N. ab E.*

1. M. *Wallichii* (N. ab E.)—*N. ab E. in Ed. n. phil. journ.* (1834) *n.* 34. *p.* 265.—Rhynchospora, *Wall. cat. n.* 3422, *a.*——Nepalia; *Wallich.*

XXIX. HAPLOSTYLIS. *N. ab E.*

1. H. *Meyenii* (N. ab E.:) involucro capitulum superante foliisque glabris, styli basi pileiformi caryopsin biconvexam tuberculo claudente.—*N. ab E. in Ed. n. phil. journ.* (1834) *n.* 34. *p.* 265; *Wight. cat. n.* 1903.—*Wall. cat. n.* 3428.—β; capitulo viviparo, coma foliorum setaceorum subsecunda.——In China; *Meyen; Vachell; Millet.* Ceylona; *Macrae*; *Wight.* In Amboina; *Lesson.*—β, Ceylona; *Macrae.*

XXX. CEPHALOSCHOENUS. *N. ab E.*

1. C. *Zeylanicus* (N. ab E.:) capitulis subtristachyis, axillaribus simpliciter corymbosis, terminalibus duplicato-corymbosis paucifloris, radiis foliisque linearibus elongatis margine scabris.—*N. ab E. in Ed. n. phil. journ.* (1834) *n.* 34. *p.* 265.——In Ceylona insula; *Macrae.*

2. C. *articulatus* (N. ab E.:) spiculis subcapitatis, capitulis fasciculatis corymbosis, corymbis axillaribus decompositis terminali supradecomposito multifloris erectis, ramulis tenuibus, bracteis inferioribus ramos æquantibus setaceis, rostro caryopsi longiore, foliis lato-linearibus carina et margine scaberrimis.—*N. ab E. in Ed. n. phil. journ.* (1834) *n.* 34. *p.* 266; *Wight. cat. n.* 1904.—Schœnus articulatus, *Roxb. fl. Ind.* 1. *p.* 184; (*ed. Wall.*) 1. *p.* 189. —S. umbellatus, *Roxb. in cœt. merc. Ind. or. mus. tab.* 703 (fide *Arn.*)——Locis uliginosis Peninsulæ regionis montanæ superioris; *Roxburgh*; *Wight.*

TRIB. V. SCLERIEÆ.

XXXI. SCLERIA. *Berg.*

Spiculæ diclines, monœcæ; fœminea uniflora: rarius spiculæ androgynæ. Squamæ distichæ aut tristichæ, in fœmineis paucæ, inferiores vacuæ. Stamina tria. Stylus trifidus. Hypogynium floris masculi monophyllum, chartaceum aut trifidum, lobis aut integerrimis aut denticulatis lacinulatisve. Nux globosa aut ovata, papillula persistente aut decidua, styli basi residua coronata, basi hypogynio persistente appresso patulove cincta.—Inflorescentia: Spiculæ vel solitariæ, et tum vero in eadem inflorescentia segregatæ, inferiores fœmineæ aut vice versa, vel geminæ, altera fœminea altera mascula, quandoque et plures glomeratæ bracteola communi propriisque suffultæ, pallidæ aut fuscidulæ: glomeruli dispositi in paniculas sæpe corymbosas, aut in spicas interruptas simplices aut compositas, axillares, versus culmi apicem magis approximatas foliisque suis decrescentibus denique in inflorescentiam terminalem, plerumque magis compositam, quandoque et simplicem atque gracilem coalescentes. Culmus sæpe altus, trigonus aut triqueter, articulatus; vaginis totis herbaceis triquetris, angulis quandoque alatis, oris margine oppositifolio, lobulo obtuso prædito. Folia angusta aut latiuscula acuminata carinata, pleraque margine serrulato-scabra.

Species orientales omnes squamis spicularum carina exalatis gaudent. Sectionis secundæ subdivisionis, " Styli basi mucroniformi atra serius decidua" insignitæ, nullæ adsunt species.

§ 1. *Spiculæ masculæ et fœmineæ in diversa parte inflorescentiæ positæ, solitariæ.*

1. S. *oryzoides* (Presl.:) culmo vaginisque acute triquetris glabris ore trilobo, panicula terminali corymbosa, spiculis solitariis ramorum ramulorumque infimis fœmineis superioribus masculis, nuce globosa lævissima, hypogynio profunde trilobo appresso.—*Presl. rel. Hænk.* 3. *p.* 201.—Scleria, *Wall. cat. n.* 3413 (ex parte).——In Luzonia; *Hænke.* In Ceylona; *Macrae.* Peninsula Ind. or.: *Wight.*

2. S. *macrophylla* (Presl.:) culmo triquetro, foliis latis lanceolatis margine scabris, panicula terminali erecta, bracteis setaceis ramulo scabro longioribus, spiculis sessilibus inferioribus fœmineis, glumis scabriusculis masculis ovatis acutis femineis ovato-subrotundis mucronatis, caryopsi globosa lævissima apiculata.——In insula Luzon; *Hænke.*

§ 2. *Spiculæ geminatæ, altera mascula, altera fœminea.*

a. *Styli basis concolor brevis persistens mucronuliformis.*

+ Hypogynium magnum membranaceo-chartaceum integrum aut repando-trilobum, nec profunde divisum.

3. S. *Sumatrensis* (Retz.:) culmo triquetro, viginis triquetris foliisque glabris, superioribus verticillato-ternis ore truncatis ciliolatis, paniculis axillaribus in terminalem ad speciem proliferam densam approximatis, ramulis divaricatis, spiculis geminis altera mascula, nucibus globosis reticulatis cinereonigris, hypogynio lato obsolete trilobo lobis patulis undatis crenatisque.—*Retz. obs.* 5. *p.* 19. *t.* 2; *Willd. sp. pl.* 4. *p.* 315; *Wight. cat. n.* 1905.—Scleria, *Wall. cat. n.* 3413 (ex parte).——In Ceylona; *Macrae.* Peninsula Indiæ orientalis; *Wight.*

4. S. *scrobiculata* (N. et Meyen:) culmo triquetro foliisque lineari attenuatis margine scaberrimis, vaginis acute triquetris scabris ore hispidulis, panicula terminali decomposita patula, spiculis alternis sessilibus bracteam setaceam subæquantibus, nuce ovata mucronulata scrobiculata.—β; paniculæ summitate abortiva, spiculis minimis pallidis.—γ; panicula tota abortiva microstachya pallida.——In Manila; *Meyen*. In insula Rawack; *Gaudichaud*.

Formæ β et γ ab abortu spicularum superiorum omnisve omnino inflorescentiæ profectæ, alieno quidem guadent habitu, propter spiculas illas duplo minores pallidasque; nec ullam tamen cum *Scl. bracteata*, Cav., habent conjunctionem. *Calamagrostis*, Rumph. amb. 6. p. 16. t. 6. 2, quam ad *Scleriam tessellatam* citat Willdenovius, *Anthistiriæ* aut saltem *Saccharinæ* cujusdam species est.

++ Hypogynium parvum, profunde trilobum aut trifidum.

5. S. *androgyna* (N. ab E.:) culmo triquetro, vaginis triquetris carinis scabris lobulo oppositifolio rotundato, foliis glabris alternis, corymbis axillaribus et terminalibus corymbosis rigidis paniculam elongatam constituentibus, spiculis aliis masculis aliis androgynis, squama tertia fœminea, quarta spiculam masculam bifloram amplectente, nucibus ovatis acutis lævibus candidis, hypogynio profunde trilobo integro.—*Wight. cat. n.* 1906.——Peninsula Indiæ orientalis; *Wight.*

Structura fere *S. flagelli*, sed notis allatis bene distincta. A *S. Sieberi* differt statura majore, perigynio non ciliato, nucibus majoribus magis acutis.

6. S. *lævis* (Retz.:) culmo triquetro vaginisque lævissimis, paniculis axillaribus corymbosis in terminalem oblongam coalescentibus, spiculis geminis altera mascula pedicellata, nucibus depresso-globosis lævissimis candidis papilla fusca, hypogynio profunde trilobo lobis integris.—*Retz. obs.* 4. *p.* 13; *Willd. sp. pl.* 4. *p.* 314.——Indiæ orientalis; *Koenig*. Ceylona; *Macrae*.

Culmo et vaginis lævigatis nucibusque depressis lævissimis facile dignoscitur.

9. S. *lithosperma* (Willd.:) culmo triquetro gracili glabro, vaginis triquetris latere oppositifolio linea media lata hirsuta, foliis linearibus angustis elongatis margine scabris, glomerulis dissite spicatis di-tristachyis, spicis axillaribus terminalique simplicibus aut inferiori composita, nuce ovata trigona lævissima basi trigono-depressa lateribus punctato-scabris, hypogynio profunde trilobo. *Willd. sp. pl.* 4. *p.* 316; *Wight. cat. n.* 1907.—S. tenuis, *Retz. obs.* 4. *p.* 13; *Willd. sp. pl.* 4. *p.* 316.—S. glaucescens, *Presl. rel. Hænk.* 3. *p.* 202.—Scleria, *Wall. cat. n.* 3416.—Scirpus lithospermus, *Lin. sp. pl.* (*ed.* 1.) *p.* 51. —*Rheed. Mal.* 12. *t.* 48.——Peninsula Indiæ orientalis. Penang; *Wallich*. Ceylona; *Macrae*.

Inter congeneres distincta nuce trigona angulis obtuse marginatis (lævissima, candida) plagis post detritum hypogynium baseos punctulis ferrugineis elevatis scabris.

8. S. *ciliaris* (N. ab E.:) culmo triquetro, foliis linearibus obtusiusculis vaginisque trialatis margine scaberrimis, lobulo oppositifolio elongato, paniculis axillaribus terminalique spicato-decompositis ovatis, bracteis basi hirtociliatis, nuce globosa, oblique arguteque undulato-ruguloso mucronata.——In China, in vicinia Macao urbis; *Vachell; Millett.*

Habitu propinqua *S. margaritiferæ*, a qua differt: nuce rugoso-exasperata concolore et bracteis basi valde ciliatis. A *S. scrobiculata* discedit: statura minore, vaginis alatis, bracteis bracteolisque ciliatis, nuce globosa, nec ovata, nec scrobiculis rugas obtusas elevantibus impressa, sed rugulis crebris undulatis interruptis acutis exasperata. *S. tessellata* minor est, gracilior, vaginis exalatis, nuce duplicato-cancellata, &c. Inter species, mihi notas, insignis est lobulo vaginarum inferiorum ovato obtuso patulo, aliquot lineas longo.

9. S. *hebecarpa* (N. ab E.:) culmo triquetro foliisque lineari-acuminatis vaginisque trialatis margine scabris, panicula terminali brevi foliacea alteraque

axillari minori ovatis, nucibus obsolete trigonis globosis mucronulatis lævibus nitidis pubescenti-hirtulis, mucrone concolore.——In Ceylona insula; *Macrae.*

Differt a *S. scabra,* Willd., nucibus non scabris, sed setulis exiguis patulis dispersis hirtulis, tum vaginis alatis. A *S. oryzoide,* Presl., nuce puberula, bracteis longioribus, communibus foliaceis, panicula non fastigiata spiculisque geminis, altera mascula; a *Scleria margaritifera* mucronulo concolore; a *S. lævi* præterea scabritie partium.

10. S. *margaritifera* (Willd.?:) culmo triquetro foliisque lineari-attenuatis margine scaberrimis, vaginis trialatis ore hispidulis, panicula terminali composita, ramis spicatis strictis alatis, spiculis subglomeratis, bracteis bracteolisque lineari-setaceis strictis, nuce lævi vertice tumido depressiusculo papillato.—*R. Br. prod. fl. nov. Holl. p.* 240; *Sieb. agrostoth. n.* 100; *Willd. sp. pl.* 4. *p.* 312; *Spr. s. v.* 3. *p.* 830.—Carex lithosperma, *Forst. prodr. n.* 339.——In insula Tanna; *Forster.* In nov. Holl.; *Brown.* In China; *Meyen.* Bahiæ in umbrosis; *Salzmann.* E Guiana allatam etiam vidi in herbario Lindl.

Vaginis alatis, ramis paniculæ bracteisque strictissimis, his decurrentibus et ramos rhachinque alatos reddentibus, tum figura caryopseos satis distincta species. Quod ad caryopsin, proxima est *Scleriæ lævi,* sed differt scabritie. *S. margaritifera,* Willd., qua " puncto obscuro apicis notata" describitur, vix hujus loci est; in nostra etenim papilla nuci concolor. *S. melaleuca,* Reichenb. in Weigelt. pl. exs. Surin., et Schlechtend. in Linnæa 6. p. 29., differt nuce globosa, subovata, nec depressa, inferne candida, mucronata, quæ in nostra minutissima papillata.

11. S. *tessellata* (Willd.:) culmo triquetro glabro gracili, vaginis triquetris latere oppositifolio strigiloso-tomentoso, spicis axillaribus terminalique subracemosis simplicibusve spiculis geminis distantibus, nuce globosa grosse cancellato-reticulata opaca.—*Willd. sp. pl.* 4. *p.* 315.—Scleria, *Wall. cat. n.* 3405. —[S. biflora, *Roxb. fl. Ind.* 3. *p.* 573; *in cœt. merc. Ind. or. mus. tab.* 1294.—Arn.]——Ceylona; *Macrae.* Peninsula Indiæ orientalis; *Koenig.* In Nepalia, et Benghalia inferiori; *Wallich.*

Nuces subglobosæ, opacæ, grosse lanceolatæ, cancellis singulis cruce humiliori divisis ideoque foveolis quatuor notatis. Vaginæ non alatæ, supremæ averso latere strigiloso-tomentosæ; culmus lævis; folia angusta, margine parum scabra.

Species dubiæ.

12. S. *poæformis* (Retz.:) culmo aphyllo, paniculæ ramis flexuosis, spiculis sessilibus, femineis axillaribus, masculis bifloris.—*Retz. obs.* 4. *p.* 13.——In India orientali.

13. S. *Zeylanica* (Poir.:) culmo triquetro scabro, foliis linearibus ligula bifida, glomerulis axillaribus paniculatis longe bracteatis, nucibus lævibus apice tuberculatis.—*Poir. enc. meth. suppl.*——Ceylona.

XXXII. CYLINDROPUS. *N. ab E.*

1. C. *junciformis* (N. ab E.)—*N. ab E. in Ed. n. phil. journ.* (1834) *n.* 34. *p.* 266.——In Ceylona insula; *Macrae.*

XXXIII. HYPOPORUM. *N. ab E.*

1. H. *pergracile* (N. ab E.:) culmo erecto filiformi simplici triquetro, glomerulis spicatis alternis paucifloris bracteam membranaceam æquantibus, nuce depresso-globosa mucronata tuberculato-echinata alba, subtus sulcis tribus eporosis impressa.—*N. ab E. in Ed. n. phil. journ.* (1834) *n.* 34. *p.* 267.—Scleria, *Wall. cat. n.* 3406.——Sillet; *Wallich.*

2. H. *capitatum* (N. ab E.:) culmo erecto simplici, spiculis capitatis foliis-

que hirsutis, nuce rugosa tuberculata basi trifariam biporosa, squamis masculis apice fimbriato-laceris.—*N. ab E. in Ed. n. phil. journ.* (1834) *n.* 34. *p.* 267.——Ceylona; *Macrae.*

XXXIV. DIPLACRUM. *R. Br.*

Spiculæ diclines, monœcæ, pauciflora. Spiculæ masculæ parvæ, ad basin fœmineæ laterales, squamis scariosis. Stamina tria. Spicula fœminea subcentralis, e squamis quatuor, quarum duæ superiores majores cuspidatæ persistentes. Hypogynium nullum aut disciforme. Stylus trifidus. Nux globosa, squamis duabus supremis spiculæ in formam utriculi bidentati conniventibus rigidulis inclusa.—Inflorescentia: Glomeruli axillares et terminales pleiostachyi. Gramina humilia. Spiculæ parvæ.

1. D. *Zeylanicum* (N. ab E.:) fasciculis tristachyis capitatis axillaribus, squamis fœmineis trifidis aristatis.—Olyra Malaccensis, *Wall. cat. n.* 3540. *b* (ex *Arn.*).——In Ceylona; *Macrae.*
An revera distinctum a *D. caricino*, R. Br.?

TRIB. VI. ELYNEÆ.

XXXV. COBRESIA. *Willd.*

1. C. *laxa* (N. ab E.:) spica subdecomposita subsecunda, spicis linearibus approximatis inferiori setaceo-bracteata composita, masculis terminalibus intra squamam nudis, squamis oblongo-lanceolatis acutis processu accessorio filiformi margine serrulato-scabro valvulam superante, culmo superne trequetro scabro, foliis lineari-acuminatis planis.—α; spica densiori, foliis latioribus culmo brevioribus.—*N. ab E. in herb. Royle. n.* 102.—β; spica laxa, spicis partialibus laxe squamosis, squamis valvula plerumque brevioribus, foliis angustioribus culmum gracilem subæquantibus.—*N. ab E. in herb. Royle, n.* 81. ——Nepalia; *Royle.*

Uncinia spartea, Spr., (quæ *Carex spartea,* Wahlenb., *C. indica,* Schkuhr.), hujus loci esse videtur. Sed seta nec evidenter clavata, nec basi ciliata, quare pro certo effari, quæ suspicor, nolebam.

XXXVI. TRILEPIS. *N. ab E.*

1. T. *Royleana* (N. ab E.:) triandra, spica composita densa, spiculis apice masculis, squamula propria subsolitaria, foliis latiusculis falcatis culmique trigoni angulis scaberrimis.—*N. et E. in herb. Royl. n.* 119; *in Ed. N. Phil. Journ.* (1834) *n.* 34. *p.* 267.——In Nepalia; *Royle.*

* T. *Lhotzkiana* (N. ab E.:) diandra, spiculis simplicibus fasciculatis axillaribus, squamulis propriis binis, foliis lineari-subulatis.—*N. ab E. in Ed. N. Phil. Journ.* (1834) *n.* 34. *p.* 267.—Caxex Lhotzkiana, *herb. Endlich.*——Brasilia, ubi in Corcovado *Lhotzky* invenit.

TRIB. VII. CARICEÆ.

XXXVII. CAREX. *Linn.*

Species Indicæ pleræque ad eam sectionem pertinent, quarum spiculæ (trigynæ) androgynæ apiceque masculæ in spicas compositas decompositasque, thyrsoideas aut corymbosas axillares pedunculatas dispositæ sunt, et in communem denique inflorescentiam terminalem confluunt. His inter nostrates tres species respondent, scil. *C. paniculata, paradoxa,* et *teretiuscula*; differunt autem itæ haud leviter eo, quod bracteis foliaceis careant omnesque spiculas circa apicem culmi in unam colligant inflorescentiam compositam brevibracteatam. Secondo loco, quod ad specierum numerum, veniunt *Carices* spiculis androgynis simplicibus aut compositis versus api-

cem culmi alternatim sessilibus, ut in *C. leporina*, *vulpina*, &c. *Carex remota* nostras etiam in Nepalia ocurrit. *C. vulpinaris vulpinæ* peraffinis est; *C. nubigena*, Don, et *notoleia*, mihi, *C. elongatam*, *virentem* et confines referunt. Haud exiguus est numerus earum, quæ spiculis gaudent solitariis axillaribus pedunculatis, aut omnibus androgynis aut infimis tantum cum aliquot terminalium masculis. Cum *C. Buxbaumii* duæ congruunt species: *C. infuscata* et *brevibracteata* mihi, quibus spica terminalis est androgyna, reliquæ autem fœmineæ. *C. lachnosperma*, Wall., et *longicruris* mihi, spiculis pedunculatis in axillis foliorum aggregatis et longe vaginatis, ad *Cobresiam* genus accedunt. *C. Davallianæ* ad latus accedit *C. trinervis* nobis. Quæ apud nos copiosissimæ eccurrunt Carices, spicis ad sexum distinctis, terminalium una aut pluribus masculis, inter Indicas nostras parcius sparguntur. Ex hisce respondet: *C. Royleana*, *Carici piluliferæ* et *tomentosæ*; *C. cæspititia*, *C. cæpitosæ*; *C. psychrophila*, *C. fuliginosæ*; *C. fusiformis*, *C. sylvaticæ*; *C. Doniana*, *C. Pseudo-Cypero*; *C. Fedia* N. ab E., et *Wallichiana*, Presc. respondent *C. pellitæ*, Muhlenb. *C. paludosa* nostas etiam in Nepalia offenditur. *C. ligularis* inter omnes excellit membrana ligulari, os vaginæ atque laminæ axillam cingente.

§ 1. *Spicula in culmo solitaria.*

1. C. *trinervis* (N. ab E.:) stigmatibus . . . , fructu . . . , spicula mascula cylindrica culmi altitudine, squamis ovalibus obtusis membranaceis brunneis linea viridi apice albo-membranaceis, foliis planis linearibus trinervibus culmoque teretiusculo lævibus.—*Royle. herb. n.* 138 (ex parte.) ——Nepalia; *Royle.*

Species distincta, sed latet fœmina; $2\frac{3}{4}$ poll. alta, spica $1\frac{1}{4}$ poll. Vaginæ longæ, striatæ. Folia 3, ad basin culmo paulo breviora, lamina angusta, dorso trinervi, demum complicata. Squamæ 3 lin. longæ, rufæ apice albæ. Stamina longa. Antheræ lineares, luteæ. Quod ad squamas *C. Davallianæ* accedit, sed sunt eæ duplo minores et folia longe diversa.

2. C. *parva* (N. ab E.:) spicula androgyna obconica, stigmatibus binis (?), fructibus conico-subulatis lævibus ore oblique truncato, squamis fœmineis inferis paucis ovatis obtusis cum acumine longo setaceo, masculis pluribus lanceolato-acuminatis, culmo striato teretiusculo folioque caulino lævibus.—*Royle. herb. n* 138 (ex parte).——Nepalia; *Royle.*

In consortio *C. trinervis* crescere videtur, quacum promiscuum unum inveni hujus speciei specimen. Inter cognitas species sola *C. Willdenoviana* accedit nostræ.

3. C. . . . . —*Royle. herb. n.* 122 (ex parte).——Nepalia; *Royle.*

Spicula longe pedunculata, tota fœminea, basi subramosa, cylindrica; ramuli 2 ad basin alterni, pauciflori. Squamæ ovatæ, mucronato-acutæ, fructus æquantes, dorso medio virides trinerves, latere rufæ. Stigmata 3: fructus ovales, subrhombei, depresso trigoni, subpedicellati, brevimucronati, tomentulo denso griseo-viridi scabro vestiti.—Adest folii pars, rigiduli, 2 lin. lati, plani, carinati, parum ad margines scabri, quod nescio an ad hanc spicam pertineat.

§ 2. *Spiculis androgynis sessilibus compositisve.*

4. C. *nubigena* (Don:) spica composita oblonga, e spiculis pluribus approximatis confertisve ovatis, bracteis e basi membranacea truncata foliaceo-cuspidatis, inferiorum foliolo spicula spicave longiore, fructibus erectis ovatis ovalibusve plano-convexis glabris dorso striatis margine nudo denticulatove, rostro brevi bidentato, stigmatibus duobus, squamis ovatis cuspidatis albido-membranaceis viridi-carinatis fructu paulo brevioribus, culmo obtuse trigono lævi, foliis lineari-angustis carinatis convolutisve margine scabris.—*Don. fl. Nep. p.* 42; *N. ab E. in herb. Royle. n.* 107, 131, 132, 133, 137, *et* 147 (ex parte); *Wight. cat. n.* 1908.—Carex, *Wall. cat. n.* 3396.——Peninsula Indiæ orientalis; *Wight.* Nepalia; *Wallich; Royle.*

Folia in nostris non convoluta, sed carinata margine scaberrima: Perigynii margines modo tuberculis aliquot aspersi, modo nudi et læves. Var. β. spiculis inferioribus compositis, et tum *Carici divulsæ* similis est, quæ quidem spiculis fructibusque duplo fere majoribus aliisque notis satis differt.

5. C. *remota* (Linn.:) spiculis androgynis basi masculis simplicibus alternis distantibus, bracteis foliaceis longissimis, stigmatibus binis, fructibus ovatis depressis glabris obtuse marginatis ore brevi obtuse bidentato, culmo trigono flaccido, foliis linearibus elongatis mollibus glabris culmum æquantibus.—*Linn. sp. pl. p.* 1383; *Willd. sp. pl.* 4. *p.* 239; *Schk. car. p.* 46. *n.* 35. *t. e. f.* 23'; *Host. gram.* 1. *p.* 40. *t.* 52; *Wahlenb. act. Holm.* 1803. *p.* 148; *N. ab. E. in herb. Royle. n.* 108, 109, 128, 134.——Nepalia; *Royle.*

Exempla Nepalensia ab Europæis differunt: spiculis (in culmo solitæ altitudinis) sæpe minoribus, apicem versus paulo confertioribus, squamis fructibusque dimidio sæpe minoribus, his magis virentibus; sed nullas præterea invenio differentias.

6. C. *notolcia* (N. ab E.:) spica basi subinterrupta, ramulis appressis dense spiculigeris oblongis, inferioribus subcylindricis longioribus, superioribus approximatis, bracteis e basi dilatata membranacea setaceis spicis suis brevioribus, infimis tantum conspicuis, fructibus erectis ovatis dorso convexo lævissimo nitente, rostro brevi bidentato marginibusque superius serrulato-scabris, stigmatibus duobus, squamis ovatis subulato-acuminatis flavo-virentibus, culmo trigono superne scabro.—*N. ab E. in herb. Royle. n.* 106. *et* 126.—Carex, *Wall. cat. n.* 3387 (ex parte).——Nepalia; *Royle; Wallich.*

Accedit *C. virenti*, Lam. et DC., sed differt spicis partialibus longioribus squamis margine non ferrugineis, dorso perigynii lævissimo. Quod ad habitum magis fere propinqua *C. divulsæ;* hujus autem fructus margine læves inveniuntur et squamæ acutiusculæ, nec longe subulatæ.

7. C. *vulpinaris* (N. ab E.:) spica ovata obtusa compacta, spiculis compositis ovatis imbricatis, fructibus ovatis bidentatis compressis subgibbosis erectis, squamis ovatis membranaceo-acutis pallide fuscis, bracteis obsoletis, culmo triquetro foliisque lineari-elongatis margine scabris.—*N. ab E. in herb. Royle. n.* 73.——Nepalia; *Royle.*

Differt a *C. vulpina* culmo graciliore, foliis angustioribus et longioribus, spica brevi ($\frac{1}{4}$–$\frac{3}{4}$ poll. longa) imbricata, squamis tenuioribus, pallide brunneis margine pallidioribus apice non herbaceo-rigiduleque cuspidatis, sed acutis acumine membranaceo, fructu dimidio fere minore. A *C. multiflora* jam bracteis minime emersis, brevissimis mere membranaceis aut brevi-setigeris, fructu majori, &c., distinguitur.

8. C. *foliosa* (Don:) spiculis pluribus androgynis compositis ovato-oblongis adpressis approximatis inferioribus remotiusculis, bracteis bracteolisque e basi lata membranacea setaceis elongatis, stigmatibus duobus, fructu ovali rostrato margine primum scabro dein glabro, squamis ovatis aristato-mucronatis testaceis carina viridi culmique triquetri angulis scaberrimis, foliis linearibus planis.—*Don. fl. Nep. p.* 42.—C. Wallichiana, *Spr. s. v. p.* 812.——Nepalia; *Wallich; Royle.*

Statura *C. muricatæ*. Adspectus spicæ fere qualis *C. intermediæ*. An *C. vulpinæ* forma?

§ 3. *Spiculis androgynis brevibus per se sessilibus, sed in spicas axillares pedunculatas solitarias vel geminas, compositas decompositasve, compositas laxasve, quandoque et corymbiformes aut disticho-plumosis dispositæ.*

9. C. *spicigera* (N. ab E.:) spiculis androgynis spicatis ovalibus quaternis superne masculis, spicis axillaribus exserte pedunculatis, stigmatibus ternis, fructibus ovatis primum compressis dein trigonis nervosis margine scabris bidentatis squamam ovatam mucronatam superantibus, foliis angustis culmo longioribus.——In Ceylona; *Macrae.*

10. C. *Lindleyana* (N. ab E.:) spicis decomposito-glomeratis ovalibus, inferioribus longe pedunculatis solitariis geminisve, supremis confluentibus, rhachi strigosa, spiculis ovatis ferrugineis apice masculis, stigmatibus ternis, fructu trigono ovato nervoso glabro rostrato squama ovata acuminato-aristata

longiore, foliis angustis apice serrulato-scabris culmo brevioribus.—*Wight. cat. n.* 1909.——In Ceylona, et Peninsula Ind. or.

Variat, α; culmi articulis superioribus infra nodos pedunculisqve strigiloso-scabris, spicis solitariis.—β; culmi articulis omnibus pedunculisque glabris angulis superius scabris, spicis mediis geminis, rachi minus strigilosa.—γ; culmo digitali, foliis breviore.

11. C. *Wightiana* (N. ab E.:) spicis decomposito-glomeratis oblongis inferioribus longe pedunculatis supremis confluentibus, rachi glabra, spiculis oblongis imbricatis apice extremo masculis, stigmatibus ternis, fructu erecto trigono ovali nervoso inter nervos scabro rostrato squama ovali acuta striata pallida duplo longiore, foliis lato-linearibus scaberrimis bracteisque culmum superantibus.—*Wight. cat. n.* 1910.——Peninsula Indiæ orientalis; *Wight.*

Proxima *C. Lindleyanæ,* sed differt: foliis longissimis latisque, spicis angustioribus oblongis, spiculis flavo-viridibus nec fusco-rufis, fructuque inter nervos valde prominulos hispido-scabro.

12. C. *raphidocarpa* (N. ab E.:) spicis compositis decompositisque oblongis densiusculis solitariis geminisve, infima longe pedunculata, reliquis subincluse pedunculatis, supremis confluentibus, rachi hirsuta, spiculis alternis approximatis ovatis paucifloris apice extremo masculis, stigmatibus ternis, fructu patente oblongo trigono longe rostrato nervoso glabro margine subciliato squama ovali subulato-acuminata duplo longiore, foliis linearibus longis subtus vaginisque hirtis.—*Wight. cat. n.* 1911.——Peninsula Indiæ orientalis; *Wight.*

Proxima hæc species *C. Lindleyanæ* et *Wightianæ* differt ab utraque jam primo adspectu fructibus squarroso-patentibus longiori rostro præditis et spicis laxioribus.

13. C. *baccans* (N. ab E.:) spicis compositis subincluse pedunculatis inferioribus remotis, rachi triquetra scabra rigida, spiculis cylindricis apice masculis inferioribus spicarum geminis ternisve, stigmatibus ternis, fructu ovato inflato dorso gibbo angulatoque ad nervos scabro brevirostri patenti-recurvo squamam ovatam setaceo-cuspidatam subæquante, bracteis omnibus foliaceis elongatis foliisque latiuscule linearibus margine nervisque scabris culmum triquetrum æquantibus.—*Wight. cat. n.* 1912.——Peninsula Indiæ orientalis; *Wight.*

Variat, β; fructu magis inflato, at minus evidenter nervoso, rugoso, crassiusculo, caryopsin non minus perfectam maturamque continente. In uno autem ejusmodi fructu alterius perigynii rudimentum ad latus caryopseos vidi.

14. C. *myosurus* (N. ab E.:) spicis compositis axillaribus exserte pedunculatis, spiculis alternis elongatis curvis multifloris superne masculis, stigmatibus ternis, fructu trigono gibbo basi attenuato rostro brevi bifido margine serrulato scabro laciniis præmorsis, squamis conformibus imbricatis ovatis margine membranaceis obtusiusculis mucronatis glabris, inferioribus quadrifariis pallidis, superioribus a medio fere masculis fuscis, foliis linearibus longis margine scabris.—*Wight. cat. n.* 1913.—Carex, *Wall. cat. n.* 3384.——Peninsula Ind. or. in montibus Dindygul (2600 pedum altitudine); *Wight.* Nepalia; *Wallich.*

A *C. eminente* vix satis differt, nisi notis declaratis, præsertim spiculis sæpe bipollicaribus inferne quadrangulis e viridi luteoque variis, apice squamis masculis arctius incumbentibus fuscisque quasi caudatis, nec multo tamen angustioribus. Bracteæ superiores filiformes, longæ.

15. C. *eminens* (N. ab E.:) spicis decompositis axillaribus exserte pedunculatis nutantibus, spiculis fasciculatis ternis quaternis attenuatis undatis multifloris superne masculis, stigmatibus ternis, fructu ovato-lanceolato hispido ciliato, rostro brevi bifido laciniis obtuse tridentatis, squamis æqualibus imbricatis brunneis margine membranaceis obtusiusculis, inferioribus fœmineis

oblongo-lanceolatis subbifariis, superioribus a medio ovatis masculis, supremis acutis, foliis linearibus longis culmoque scabris.—*Carex*, *Wall. cat. n.* 3397.——Gossain-than; *Wallich.*

16. C. *Indica* (Willd.:) spicis subdecompositis subpinnatis axillaribus inferioribus exserte pedunculatis, rachi angulata scaberrima, spiculis alternis subdistichis teretibus acutis basi fœmineis, stigmatibus ternis, fructu trigono-ovato rostrato nervoso striato scabro, squamis subæqualibus imbricatis ovatis mucronatis fœmineis paulo longius mucronatis scabris, foliis latiuscule linearibus margine scabris, floralibus spicas suas superantibus.—*Willd. sp. pl.* 4. *p.* 246; *Wahlenb. act. Holm.* 1803. *p.* 149; *N. ab E. in herb. Royle. n.* 23. *et* 85; *Wight. cat. n.* 1914.—Carex, *Wall. cat. n.* 3400. *a* (ex parte), *b* (ex parte).——Nepalia; *Royle*; *Wallich.* Peninsula Ind. or.; *Wight.*

Similis *C. meiogynæ*, sed facilis distinctu e spiculis basi non laxe squamosis propterea que non attenuato-sessilibus, tum squamis conformibus ovatis nervo in mucronem producto, quandoque autem submuticis, fœmineis micanti-scabris dorso viridibus margine pallide lutescentibus.

17. C. *condensata* (N. ab E.:) spicis decompositis densis ovatis axillaribus, inferioribus exserte pedunculatis rhachi angulata sulphureo-hispidissima, spiculis alternatim approximatis lanceolatis teretibus acutis basi fœmineis, stigmatibus ternis, fructu ovato-lanceolato rostrato basi trigono, squamis æqualibus imbricatis ovatis mucronatis membranaceis griseis, fœmineis longius mucronatis, foliis latiuscule linearibus margine culmoque trigono scabris, floralibus spicas suas superantibus margine hispidulis.—*N. ab E. in herb. Royle. n.* 83.—Carex, *Wall. cat. n.* 3400. *b* (ex parte).—β, *flava*; spiculis flavescentibus, inflorescentia breviori (nondum evoluta).—Carex, *Wall. cat. n.* 3398.——In Nepalia; *Royle.* Kamaon; *Wallich.*

18. C. *valida* (N. ab E.:) spicis fasciculato-decompositis androgynis pedunculatis solitariis, inferioribus axillaribus remotis, terminalibus corymbosis, spiculis subglomeratis inferne patulo-fœmineis superne masculis, stigmatibus ternis, fructibus ovatis trigonis rostratis patulis glabris squama ovata acuminata membranacea fusco-purpurascente paulo longioribus, culmo crasso valido, involucris foliisque lato-linearibus glabris rigidis culmum æquantibus superantibusve.——In Ceylona. China, ad Macao et in insulis vicinis; *Vachell.*

19. C. *meiogyna* (N. ab E.:) spicis compositis axillaribus, inferioribus exserte pedunculatis, rachi angulata rachillaque scabra, spiculis alternis approximatis teretibus acutis basi laxe squamosis fœmineis, stigmatibus ternis, fructibus . . . , squamis fœmineis minoribus ovato-lanceolatis subacutis, masculis pluribus arcte imbricatis ovatis obtusis brunneis, foliis latiuscule linearibus planis margine scaberrimis floralibus plerisque spicas suas superantibus.—*N. ab E. in herb. Royle. n.* 82; *Wight. cat. n.* 1915.—β; spiculis minus patulis, subimbricatis.—*Wight. cat. n.* 1915. *b.*—Carex, *Wall. cat. n.* 3400. *e.*——Nepalia; *Wallich*; *Royle.* Peninsula Ind. or.; *Wight.*

Differt a *C. ramosa*: spiculis angustioribus basi laxe squamosis et squamis masculis obtusis, nec fœmineis evidenter acutatis, omnibus partibus minoribus et angustioribus.

20. C. *filicina* (N. ab E.:) spicis pinnato-decompositis supradecompositisque androgynis pedunculatis densis, spiculis superne masculis inferne laxe fœmineis, stigmatibus ternis, fructibus lanceolatis subulato-rostratis squamis ovatis obtusiusculis longioribus, culmo laxo, foliis latis scabris.—*Wight. cat. n.* 1916.—β, *minor*; spicis minoribus, magis approximatis, fructibus paulo brevioribus.—*Wight. cat. n.* 1916. *b.*——In Peninsula Indiæ orientalis; *Wight.* Ceylona; *Macrae.*

Pulchra spicies, *C. polystachyæ*, magisque etiam *cladostachyæ*, Wahlenb. similis, sed diversa dispositione pinnata et colore rubro-fusco spicularum, foliis 4–5 lineas latis, squamis ovatis, tempore fructus obtusis.

21. C. *cruciata* (Wahlenb.:) spicis subdecompositis axillaribus infima ex-

serte pedunculata, rachi angulata scabra, spiculis alternis divaricatis distantibus linearibus laxe squamosis, squamis masculis terminalibus subternis, fœmineis pluribus, stigmatibus ternis, fructu subrecurvo oblongo subtrigono nervoso-striato glabro longirostri, squamis fœmineis alternis ex ovata basi setaceis setaceove aristatis fructu brevioribus scabris, foliis latiuscule linearibus culmum superantibus culmoque gracili trigono lævibus, foliis floralibus spicas suas superantibus.—α; squamis pallidis longe setaceis.—*N. ab E. in herb. Royle. n.* 86 *et* 87.—C. cruciata, *Wahlenb. act. Holm.* 1803. *p.* 149; *Willd. sp. pl.* 4. *p.* 248.—Carex, *Wall. cat. n.* 3400. *a* (ex parte).—β; squamis pallide ferrugineis ovatis setaceo-aristatis, spicis communibus ternis brevioribus ovato ambitu, decompositis densioribus.—*Wight. cat. n.* 1917.——α. Nepalia; *Royle.* Kamaon; *Wallich.*—β. Peninsula Ind. or.; *Wight.*

A *C. filicina* differt inflorescentia laxiore et squamis subulato-cuspidatis.

§ 4. *Spiculis omnibus androgynis solitariis aut compositis, pedunculis axillaribus fasciculatis.*

22. C. *lachnosperma* (Wall.:) spicis filiformibus cernuis apice masculis longe pedunculatis, radiis fasciculatis inferioribus terminalibusque solitariis, bracteis foliaceis elongatis, vaginis pedunculo multo brevioribus, fructibus ovatis nervoso-striatis pubescenti-scabris patulis, squamis ovalibus mucronato-acutis squama brevioribus margine scabris elongatis.—*Wall. cat n.* 3379; *N. ab E. in herb. Royle. n.* 103.—C. flexilis, *Don. fl. Nep. p.* 42.—C. lenta, *Spr. s. v.* 3. *p.* 811.——Nepalia; *Royle*; *Wallich.*

23. C. *longicruris* (N. ab E.:) spiculis fasciculatis incluse pedunculatis compositis simplicibusve laxis incluse pedunculatis apice masculis, vaginis brctealibus superioribus partialibusque setigeris, stigmatibus binis capillaribus prælongis, fructibus ovalibus biconvexis rostratis glabris, junioribus margine ciliato-serratis, squamis fœmineis alternis patulis ovali-oblongis mucronato-acutis luteo-rufis, culmo gracili, foliis longis linearibus strictis supra scaberrimis.—*N. ab E. in herb. Royle. n.* 93; *Wight. cat. n.* 1918.——Nepalia; *Royle.* Peninsula Ind. or.; *Wight.*

C. lachnospermæ inter cognitas maxime propinqua.

§ 5. *Spiculis omnibus androgynis pedunculatis, superioribus solitariis, infimis quandoque compositis.*

24. C. *longipes* (Don.:) spicis pluribus alternis anguste cylindricis laxis apice masculis, inferioribus basi quandoque compositis pedunculo longe exserto, stigmatibus binis elongatis, fructibus (maturis) late ovalibus glabris nervoso-striatis longe rostratis, rostro margine serrulato scabro apice bifido, squamis oblongis aristatis flavis dorso viridi fructum æquantibus, culmo gracili, foliis angustis scabris brctealibusque elongatis.—*Don. fl. Nep. p.* 42; *Wight. cat. n.* 1919.——In Nepaliæ alpibus; *Wallich.* Peninsula Indiæ orientalis; *Wight.*

Simillima *C. lachnospermæ*, Wall., a qua differt fructibus magis depressis latioribus glabris spicisque inferioribus basi quandoque compositis, (licet una alterave tantummodo adsit spicula partialis, in aliis autem nulla omnino); tum squamis in acumen setaceum squama paulo brevius excurrentibus. Pedunculi gracillimi, hinc spiculæ sæpe nutantes. *C. longicruris* differt: spicis incluse pedunculatis paucifloris, squamis brevimucronatis, fructibus minoribus magis convexis.

§ 6. *Spiculis omnibus androgynis, solitariis, pedunculatis.*

25. C. *asperula* (N. ab E.:) spicis subquinis pedunculatis cylindricis, terminali inferne mascula, infima remotiuscula, bracteis foliaceis evaginatis, stigmatibus ternis, fructibus ovalibus depressis quinque-costatis subtilissime punctato-exasperatis, rostro angusto bidente fructu paulo breviori, squamis

fœmineis oblongis acutis, masculis lanceolato-subulatis, culmo triquetro angulis scabris.—*N. ab E. in herb. Royle. n.* 111.——Nepalia; *Royle.*

C. Caucasicæ, Steven. et *Buxbaumii* nonnihil affinis, differt rostro fructus angusto, ejusque superficie dense punctulato-aspera et basin versus nervoso-plicata, 5–7-costata.

26. C. *hæmatostoma* (N. ab E.:) spiculis subsenis alternis exserte pedunculatis erectis cylindricis basi masculis (an dioicis?) inferioribus bracteatis, stigmatibus ternis, fructibus oblongo-lanceolatis pubescentibus ore fusco-sanguineo bidentatis, squamis ovatis brevi-cuspidatis fusco nitidis albo-carinatis marginatisque, culmo compresso striato inferne lævi, foliis linearibus planis margine et carina scabris culmum æquantibus.—*N. ab E. in herb. Royle. n.* 116.——Nepalia; *Royle.*

In nostro exemplo spicæ omnes pure fœmineæ squamis paucis ad basin omnino sterilibus reliquis conformibus. Non sufficit igitur hoc unum specimen ad rite stabiliendam speciem. Attamen distinctæ speciei esse in propatulo est. Ad *C. atratam* quodammodo accedit.

27. C. *Nepalensis* (Spr.:) digyna, vaginis elongatis pedunculo brevioribus, spicis strictis cylindraceis apice masculis, glumis lanceolatis longe cuspidatis, fructibus ovatis rostratis scaberimis costatis apice bipartitis.—C. macrolepis, *Don. fl. Nep. p.* 42.——In Nepaliæ alpibus.; *Wallich.*

Inter exempla, quæ vidi, hæc non occurrebat.

§ 7. *Spicula terminali androgyna, reliquis fœmineis.*

28. C. *infuscata* (N. ab E.:) spicis 4–5 valde approximatis ovalibus sessilibus, mediis minoribus, suprema basi mascula, inferioribus fœmineis infimave sola foliaceo-bracteatis, bractea evaginata, stigmatibus ternis, fructibus ovalibus gibboso-trigonis glabris mucrone brevi truncato squamas ovatas acutas fuligineo-nigras superantibus, culmo triquetro apice scabro incurvo.—*N. ab E. in herb. Royle. n.* 118. *et* 123.—? β; microcarpa.—Carex, *Wall. cat. n.* 3381 (ex parte, specimen mutilum).——Nepalia; *Royle; Wallich.*

Differt a *C. Buxbaumii:* culmo humiliore incurvo, foliis culmo brevioribus, spicis pluribus brevioribus omnibus aut superioribus saltem præter infimam in formam fere capituli confertis, mediisque fere duplo minoribus basi supremæ appressis, squamis fere concoloribus fusco-nigris, nervo carinali in paucissimis apicem versus virente, in plerisque paulo pallidiore, fructu angustiore brevi mucrone prædito nec omnino truncato, squamis masculis non majoribus, ut in *C. Buxbaumii*, sed minoribus reliquis. Var. β. fortasse diversæ est speciei, differt squamis fructibusque dimidio minoribus, brevioribus, illis obtusiusculis, culmo altiore gracili.

29. C. *parvibracteata* (N. ab E.:) spicis fœmineis ternis, inferiori remota pedunculata reliquis subsessilibus, terminali inferne masculo, bracteis evaginatis subulatis brevissimis, stigmatibus ternis, fructibus oblongis apice emarginato-bidentatis stylo prominulo, squamis ovatis acutiusculis apice denticulatis fuscis carina viridi, masculis majoribus, culmo triquetro gracili angulis scabris, foliis elongatis.—*N. ab E. in herb. Royle. n.* 124.——Nepalia; *Royle.*

Est e societate *C. Buxbaumii, asperulæ* et affinium. Squamæ acutæ, nec acuminatæ, apice denticulatæ et scabriusculæ, fusco-sanguineæ, margine albo, carina viridi. Stylus exsextus. Culmus gracilis, acute triqueter, angulis scabris. Folia plana, acuminata, latiuscula, scabra, culmo altiora. Bracteæ absque vagina, lineari-subulatæ, spica sua multo breviores. Spicæ tres, inferior remotior, pedunculata, reliquæ sessiles.

§ 8. *Spiculis solitariis, inferioribus pedunculatis androgynis, suprema aut sola aut pluribus masculis.*

30. C. *concolor* (N. ab E.:) spiculis subternis maxime distantibus, una subradicali unaque media androgynis linearibus superne masculis longe pe-

dunculatis brevivaginatis, suprema (an etiam androgyna?) nuda, stigmatibus ternis, fructibus grossis ventricoso-trigonis rostratis dense subtiliterque striatis carinis scabris squamis rhombeo-ovatis obtusis confertim striatis totis herbaceo-viridibus duplo longioribus, culmo trigono, foliis latiusule linearibus planis acuminatissimis culmo longioribus.—Carex, *Wall. cat. n.* 3391.——Rajmahl; *Wallich.*

Planta speciosa, elatior, foliis multis fere radicalibus longis insignis.

31. C. *phacota* (Spr.:) spicis cylindricis subquinis approximatis nutantibus, inferioribus fœmineis apice masculis, terminali binisve terminalibus masculis, bracteis foliaceis evaginatis, stigmatibus binis, fructibus obovatis basi cuneatis brevissime rostratis papilloso-micantibus depressis, squamis obcordatis e sinu setaceo-aristatis, arista feminea squama fructuque longiore, mascula breviore, culmo angustissime triquetro, foliis bracteisque culmum superantibus linearibus planis margine carinaque scaberrimis.—*Spr. s. v.* 3. *p.* 826; *Wight. cat. n.* 1920.—C. lenticularis, *Don. prod. fl. Nep. p.* 43.——In Nepaliæ alpibus; *Wallich.* Peninsula Ind. or.; *Wight.*

Cl. Don flores masculos in apice spicularum fertilium haud observavit. Squamæ masculæ castaneæ, carina viridi; fœmineæ virides, margine castaneo pallidiori.

32. C. *hymenolepis* (N. ab E.:) spicis cylindricis (subquinis) alternis, inferioribus androgynis superne masculis incluse pedunculatis, superioribus sessilibus totis masculis, stigmatibus ternis, fructibus (immuturis) rostratis bidentatis scabris (lanceolatis), squamis oblongo-ellipticis albo-membranaceis subulato-cuspidatis, culmo triquetro, foliis linearibus longissimis scabris, radice repente.—Carex, *Wall. cat. n.* 3385.——Nepalia; *Wallich.*

Affinis *C. setigeræ*, at major, et spicis albis distincta.

33. C. *setigera* (Don.:) spicis pluribus cylindricis erectis, inferioribus androgynis pedunculatis pedunculo infimæ remotiusculo exserto, superioribus 2–3 approximatis sessilibus masculis, stigmatibus ternis, squamis membranaceis late ovalibus cuspidato-mucronatis testaceis limbo pallido, fructu late ovato brevirostri pubescenti-scabro; bractea inferiori vaginante, foliis linearibus margine carinaque scabris.—*Don. fl. Nep. p.* 43.—Carex, *Wall. cat. n.* 3381 (ex parte).—β, *ustilaginosa.*—*N. ab E. in herb. Royle. n.* 117.—γ, *humilis;* culmo 2–3-pollicari, foliis magis glaucis, complicatis.—*N. ab E. in herb. Royle. n.* 121.——Nepala; *Wallich; Royle.*

§ 9. *Spicis sexu distinctis in eodem culmo, mascula singula terminali.*

a. *Spicis fœmineis sessilibus vel incluse pedunculatis.*

34. C. *obscura* (N. ab E.:) spicis fœmineis pluribus imbricatis lanceolatis, inferiori pedunculata, reliquis sessilibus subsessilibusve, terminali mascula obovato-lanceolata, bracteis foliaceis evaginatis, stigmatibus ternis, fructibus (immaturis) oblongis emarginato-bidentatis margine scabris, squamis ovatis obtusis fuscis, culmi triquetri angulis foliisque planis margine scaberrimis.—*N. ab E. in herb. Royle. n.* 112.——Nepalia; *Royle.*

Ad *C. alopecurum*, Don., et *C. punctatam* accedit.

35. C. *Victorialis* (N. ab E.:) spica mascula una, fœmineis subternis incluse pedunculatis erectis cylindraceis angustis, stigmatibus ternis, fructibus lanceolatis (?) rostratis ore integro glabris, squamis fœmineis late ovalibus emarginatis interjecto mucrone, masculis majoribus elliptico-oblongis subrepandis obtusis cum mucrone, culmo humili, foliis bracteisque foliaceis linearibus vaginatis margine scaberrimis, rhizomate crasso fusco-fibroso.—*N. ab E. in herb. Royle. n.* 145.——Nepalia; *Royle.*

Exemplum nostrum floribus gaudet absque fructu; distinctam tamen esse speciem apparet. *C. miliaceæ*, Willd., verbis descriptionis magis, quam reapse convenit.

36. C. *Royleana* (N. ab E.:) spica mascula solitaria, fœmineis subternis cylindricis, binis superioribus approximatis sessilibus, inferiori incluse pedunculata, pedunculo brevi, bracteis foliaceis subevaginatis, stigmatibus ternis, fructibus ovatis subrostratis bidentatis pubescentibus squamam ovalem subulato-cuspidatam æquantibus, culmo gracili triquetro basi folioso.—α; spicis fœmineis duabus.—*N. ab E. in herb. Royle. n.* 36.—β; spicis fœmineis tribus.—*N. ab E. in herb. Royle. n.* 152.——Nepalia; *Royle.*

Vices gerit, inter *Carices* Nepalenses tum *C. piluliferæ*, tum *C. tomentosæ*. Differt autem a cunctis affinibus: squamis pallidis viridulis rigidulisque in cuspidem subulatum longitudine ipsius squamæ productis et spica mascula lutea, squamis etiam acutis, apice nonnihil patulis.

b. *Spicis fœmineis magis minusve exserte pedunculatis.*

37. C. *ligulata* (N. ab E.:) spica mascula solitaria, fœmineis subquinis approximatis subexserte pedunculatis erectis anguste cylindricis, stigmatibus ternis, fructibus oblongis trigono-gibbis rostratis pubescenti-asperulis binervibus, squamis ovatis fœmineis ex apice obtusiusculo breviaristatis pallidis, masculis muticis, culmo trigono ad apicem folioso, foliis bracteisque vaginatis ligulatisque planis culmo longioribus.—*N. ab E. in herb. Royle. n.* 115; *Wight. cat. n.* 1921.——Nepalia; *Royle.* Peninsula Ind. or.; *Wight.*

A Caricibus omnibus distincta vaginis culmum totum vestientibus, altero latere apicem versus ferrugineo-membranaceis, hac membrana ad basin laminæ transeunte apiceque libera ligulam veram obtusam formante, inferioribus hispido-scabris.

38. C. *punctata* (N. ab E.:) spica mascula solitaria, fœmineis ternis subcylindricis basi attenuatis approximatis subfastigiatis, inferioribus pudunculatis; superiori subsessili, bracteis setaceis evaginatis auriculatis, stigmatibus binis stylo exserto, fructibus ovalibus convexis brevirostribus quinquenervibus glabris sanguineo-punctatis squamis ovatis obtusis longioribus, culmo triquetro angulis foliisque planis rigidis margine scaberrimis.—*N. ab E. in herb. Royle. n.* 110 (squamis masc. fuscis), *et* 125 (sq. masc. pallidis).——Nepalia; *Royle.*

Ob puncta fructuum haud absimilis *C. lenticulari*, Don., sed differt squamis obtusis et fructus figura, maxime autem spiculis inferioribus haud androgynis (quod semper in *lenticulari* videbis, etiamsi cl. Don. non observaverit). Squamæ sunt fuscæ, linea media lata virente. An tamen *C. lenticularis* var.?

39. C. *cardiolepis* (N. ab E.:) spica mascula solitaria basi bracteata, fœmineis duabus distantibus pedunculatis laxe cylindricis, pedunculo semivaginato vagina ore membranacea brachyphylla, stylo e perigynio prominente, stigmatibus ternis, fructu ovali pubescenti-scabro ore oblique truncato, squamis ovalibus emarginatis e sinu subulatis, fœmineis margine candidis, masculis totis fuscis, culmo obsolete trigono foliisque linearibus culmum æquantibus scaberrimis.—*N. ab E. in herb. Royle. n.* 146.——Nepalia; *Royle.*

Distinctissima squamis late emarginatis e sinu subulato-mucronatis, fœmineis dorso fuscis margine late margaritaceis membranaceis, masculis fuscis.

40. C. *cæspititia* (N. ab E.:) spica masc. una oblongo-cylindracea, fœmineis ternis subapproximatis minoribus appressis cylindricis, bracteis foliaceis evaginatis culmi triquetri angulis foliisque margine serrulato-scaberrimis, infima remotiuscula, stigmatibus binis, perigynio glabro obtuso ore integro, squamis ovato-oblongis obtusis, fœmineis minoribus viridibus, masculis utrinque vitta ferruginea notatis.—Carex, *Wall. cat. n.* 3392.——Silhet; *Wallich.*

Habitu ad *C. cæspitosam* accedit, sed differt characteribus adnotatis.

41. C. *psychrophila* (N. ab E.:) spica mascula unica, fœmineis ternis, inferiori longe pedunculata cernua, superioribus subsessilibus, bracteis foliaceis vaginis subnullis, stigmatibus ternis, fructibus ovalibus trigonis rostratis glabris squamam oblongam acutam nigram superantibus, culmo acute triquetro

scabro, foliis planis culmo brevioribus.—*N. ab E. in herb. Royle. n.* 113.——Nepalia; *Royle.*

Similis primo adspectu *C. fuliginosæ* et *frigidæ*, sed differt statura altiore et firmiore, bracteis evaginatis solaque infima vagina brevissima instructa, fructu breviori gibboso abrupte rostrato.

42. C. *fusiformis* (N. ab E.:) spicis masculis subbinis, fœmineis quaternis remotis filiformibus laxis erectis (?), bracteis foliaceis vaginatis, stigmatibus ternis, fructibus fusiformibus attenuato-rostratis squama ovali mucronulata membranacea duplo longioribus, ore membranaceo bidentato, culmo elongato gracili, foliis planis margine scabriusculis.—*N. ab E. in herb. Royle. n.* 88.——Nepalia; *Royle.*

Ad *C. sylvaticam* nostratem accedit proxime, differt autem evidenter squamis non lanceolatis attenuatis, sed ovalibus, planis, brevi mucronulo præditis, in ambitu albis, in disco pallide viridibus, et fructibus angustioribus, utrinque, maxime autem superius attenuatis, rostri apice albo-membranaceo vidente. Habitus omnino *C. sylvaticæ,* et spicæ fœmineæ probabiliter etiam nutant.

43. C. *Doniana* (Spr.:) spiculis fœmineis ternis quaternisve erectiusculis cylindricis brevipedunculatis, mascula solitaria, bracteis inferioribus foliaceis evaginatis, fructu patulo ovato lanceolato margine acuto in rostrum planum bidentatum attenuato obsolete nervoso glabro, squamis e basi latiori lanceolato-subulatis scabris albo membranaceis angustis nervo carinali viridi excurrente fructu brevioribus, foliis linearibus planis margine et culmo triquetro angulis scabris.—*Spr. s. v. 3. p.* 825.—C. chlorostachys, *Don. fl. Nep. p.* 43.—Carex, *Wall. cat. n.* 3395.——Nepalia et Sillet; *Wallich.*

Est inter *Carices* Indicas tanquam *C. Pseudo-cyperi* nostratis æmula.

44. C. *alopecuroides* (Don.:) trigyna, vaginis nullis, spicis fœmineis erectis cylindraceis subsessilibus, masculis solitariis, glumis ellipticis acuminatis superne scabris, fructibus lanceolatis compressis lævibus apice truncatis emarginatis.—*Don. fl. Nep. p.* 43.—In Nepaliæ alpibus; *Wallich.*

45. C. *Retzii* (N. ab E.:) spicis erectis, mascula terminali, fœmineis pedunculatis quaternis, fructibus acuminatis.—C. chinensis, *Retz. obs.* 3. *p.* 42.——China; *Bladh.*

§ 10. *Spicis sexu distinctis in eadem stirpe masculis pluribus.*

46. C. *tenuis* (N. ab E.:) spicis masculis ternis sessilibus, infera bractea membranacea setigera suffulta, fœminea solitaria remotiuscula lanceolata incluse brevipedunculata, bractea setacea, stigmatibus ternis, fructibus lanceolatis compressis margine scabris ore truncato edentulo, squamis ovatis acutiusculis membranaceis rufo-ferrugineis albo-marginatis, culmo gracili compresso basi folioso foliisque angustis linearibus scabris, foliis radicalibus culmum superantibus.—*N. ab E. in herb. Royle. n.* 120, 122.——Nepalia; *Royle.*

47. C. *cruenta* (N. ab E.:) spicis masculis subcylindricis nutantibus fœmineisque subternis oblongis pendulis longe pedunculatis distantibus, bracteis vaginatis foliaceis decrescentibus, stigmatibus ternis, fructibus oblongo-lanceolatis acuminatis bidentibus membranaceis atro-sanguineis glabris squamis lanceolatis aristato-cuspidatis concoloribus duplo longioribus, caryopsi longe pedicellata trigona, culmo gracili compresso-trigono striato lævi.—Carex, *Wall. cat. n.* 3389.——Kamaon; *Wallich.*

Primo intuitu *C. frigidam* fingit aut *fuliginosam.* Distincta autem est præ omnibus perigynio semper molli et membranaceo, caryopsi intra perigynium pedicello longitudine fere caryopseos suffulta, colore atro-sanguineo spicarum fœminearum, masculis spicis ternis pedunculatis bracteisque vaginalibus dehiscentibus microphyllis stipatis cernuis.

48. C. *Fedia* (N. ab E.:) spicis masculis fœmineisque ternis his cylindricis distantibus erectis inferiori exserte pedunculata, bracteis foliaceis inferiori vaginante, stigmatibus ternis, fructibus ovato-ventricosis brevirostribus hirtis apice obliquis bi-trilobis squamam ovalem cuspidatam subæquantibus sed duplo latioribus, culmo triquetro lævi, foliis planis margine scabris.—*N. ab E. in herb. Royle. n.* 114.——Nepalia; *Royle.*

A *C. pellita*, Muhlenb., ægre distinguitur fructibus majoribus ventricosis, ore non tam bidentatis dentibus acutis, quam bi-etiam tri-quadrilobis, lobis lanceolatis planiusculis serrulatis obliquis ita ut *Fediæ* fructum haud male repræsentet. Reliqui autem characteres omnes fere in *Caricem pellitam* congruunt. A *Carice Wallichiana* distinguendam hanc esse censebam ob illius spicas fœm. duas magis a masculis distantibus et omnino sessilibus, cum nostræ tres sint, quarum inferior pedunculata vaginaque bracteæ suæ, satis longa, ad medium pedunculum vestita. An var.?

49. C. *Wallichiana* (Presc.:) spicis masc. ternis, fœmineis binis remotis subsessilibus erectis cylindricis, bracteis foliaceis subevaginatis longis, stigmatibus ternis, fructu . . . . ., perigynio floris valide bidentato hirto, squamis oblongo-lanceolatis acuminatis subaristatis pallide viridibus, masculis latioribus repandis fusco-testaceis membranaceis.—*Prescott in Wall. cat. n.* 3380. ——Oude, Torraye; *Wallich.*

50. C. *paludosa* (Good.:) spiculis masculis binis ternisque, fœmineis subternis elongato-cylindricis erectis, inferioribus pedunculatis, bracteis foliaceis evaginatis, stigmatibus ternis, fructu ovato nervoso bidentato glabro squama lanceolata aristato-cuspidata latiore eandemque æquante, culmi triquetri angulis foliisque elongatis carinatis margine scabris.—*Good. act. soc. Linn. Lond.* 2. *p.* 202; *Willd. sp. pl.* 4. *p.* 305; *Wahlenb. act. Holm.* 1803, *p.* 163; *Schk. car. p.* 121. *n.* 101. *t. Oo. et Vv. f.* 101; *Host. gr. Austr.* 1. *p.* 68. *t.* 92; *Hoppe car. p.* 94. *n.* 103; *N. ab E. in herb. Royle. n.* 130.——Nepalia; *Royle.*

Forma est squamis fœm. minus acutis solito majoribus $2\frac{3}{4}$ lin. longis, qualem e Flora Brasiliensi et Erlangensi omnino similem coram habeo.

XXXVIII. UNCINIA. *Pers.*

1. U. *Nepalensis* (N. ab E.:) spica solitaria stricta multiflora superne mascula, fructibus lanceolatis compresssis obtusis bidentatis, arista subcylindrica scabra inclusa foliis filiformibus.—*N. ab E. in herb. Royle. n.* 147 (ex parte), *et* 309.——Nepalia; *Royle.*

Distincta culmo humili arista, crassa inclusa germine paulo longiore, apicis appendiculo brevi pallido filiformi bilobo inflexa, (humili rudimento).

NOTE.

In performing the office of an Editor, I have not altered the locality assigned by the author to Royle's plants: at the same time it is right to state, that none, or almost none, were obtained in the valley of Nepal or adjacent mountains; but that Mr Royle collected almost all either in the Himalaya range or in the north-west parts of Bengal. The particular stations to the respective numbers will be given ere long by Mr Royle himself.—*G. A. W.-A.*

INDEX

OF THE

ORDERS, TRIBES, GENERA, AND SPECIES.

PRINTED BY NEILL & CO. OLD FISHMARKET.